The Rise of
Systems Theory

The Rise of Systems Theory

AN IDEOLOGICAL ANALYSIS

ROBERT LILIENFELD

A WILEY-INTERSCIENCE PUBLICATION

JOHN WILEY & SONS   New York · Chichester · Brisbane · Toronto

Library of Congress Cataloging in Publication Data:
Lilienfeld, Robert.
  The rise of systems theory.

  "A Wiley-Interscience publication."
  Includes bibliographical references and index.
  1. System theory—History. 2. System analysis. 3. United States—Social policy. I. T
HN59.L53   309   77-12609
ISBN 0-471-53533-8
Printed in the United States of America
1 0 9 8 7 6 5 4 3

Acknowledgments

I thank the following authors and publishers for permission to quote from copyrighted works:

Quotation, p. 149: from *Game Theory: A Nontechnical Introduction,* by Morton D. Davis, pp. 16–18, © 1970 by Morton D. Davis, Basic Books, Inc., Publishers, New York.

Quotations, pp. 120, 121–122, 123–124, 125–126, 127, 128, 192, 233: from *Systems Analysis in Public Policy: A Critique,* by Ida R. Hoos, © 1972 by The Regents of the University of California, reprinted by permission of the University of California Press, Berkeley.

Quotations, pp. 190, 195: from *The Sciences and the Humanities: Conflict and Reconciliation,* by William T. Jones, © 1965 by The Regents of the University of California, reprinted by permission of the University of California Press, Berkeley.

Quotation, p. 269: from *The Administrative State,* by F. Morstein Marx, published by the University of Chicago Press, Chicago, © 1957.

Quotations, pp. 161, 162, 167, 168–169, 171, 172, 176: from *Introduction to Systems Philosophy,* by Ervin Laszlo, published by Gordon & Breach, New York, 1972.

Quotations, pp. 250–251, 251–252, 252–253, 253–254, 254–255: from *Physics and Philosophy,* by Werner Heisenberg, © 1958 by Werner Heisenberg, reprinted by permission of Harper & Row, Publishers, New York.

Quotations, pp. 212, 213–214, 215: from *The Nerves of Government,* by Karl Deutsch, reprinted by permission of Macmillan Publishing Co., Inc., © 1963, 1966 by The Free Press, a division of the Macmillan Company, New York.

Quotations, pp. 271, 272, 273, 274, 275–276: from *The Elite in the Welfare State,* by Piet Thoenes, reprinted with permission of Macmillan Publishing Co., Inc., © 1966 by The Free Press, a division of Macmillan Company, New York. Printed in Great Britain by Faber & Faber, Ltd., London.

Quotation, pp. 190–191: from *Methodology of the Social Sciences,* by Max Weber, reprinted with permission of Macmillan Publishing Company, Inc., © 1949 by The Free Press, a division of the Macmillan Company, New York.

Quotations, pp. 115–116, 118: from *Systems Analysis: A Diagnostic Approach,* by Van Court Hare, 1967, reprinted by permission of Harcourt Brace Jovanovich, New York.

Quotation, p. 78: from "Art for Art's Sake," in *Two Cheers for Democracy,* by E.M. Forster, 1951, reprinted by permission of Harcourt Brace Jovanovich, New York.

Quotations, pp. 260, 260–261, 261–262: from *Ideology and Utopia*, by Karl Mannheim, 1965, reprinted by permission of Harcourt Brace Jovanovich, New York.

Quotations, pp. 66, 67, 68–69, 71–72, 73, 74: from *The Human Use of Human Beings*, by Norbert Wiener, reprinted by permission of Houghton Mifflin Company, Boston,

Quotations, pp. 235, 236, 237: from *Urban Dynamics*, by Jay Forrester, reprinted by permission of the M.I.T. Press, Cambridge, Massachusetts, © 1969, The M.I.T. Press.

Quotations, pp. 179–180, 181, 182, 183, 184, 187: from *Concept and Quality*, by Stephen Pepper, 1967, reprinted by permission of the Open Court Publishing Company, LaSalle, Illinois.

Quotation, pp. 143–144: from *Essays in Economics*, Vol. 1, 2nd ed., by Wassily Leontief, International Arts and Sciences Press, New York, 1976, reprinted by permission of the author.

Quotation, p. 175: from "Science as a Vocation," by Max Weber, in *From Max Weber*, by Gerth and Mills, Oxford University Press, New York, 1958.

Quotations, pp. 197–198: from *Societies—Evolutionary and Comparative Perspectives*, by Talcott Parsons, © 1966, reprinted by permission of Prentice-Hall, Inc., Englewood Cliffs, New Jersey.

Quotations, pp. 198–199: from *The System of Modern Societies*, by Talcott Parsons, © 1971, reprinted by permission of Prentice-Hall, Inc., Englewood Cliffs, New Jersey.

Quotation, pp. 201, 202, 203, 204–205: from *Sociology and Modern Systems Theory*, by Walter Buckley, © 1967, reprinted by permission of Prentice-Hall, Inc., Englewood Cliffs, New Jersey.

Quotation, pp. 258–259: from "The Concept of Ideology" in *History and Theory* 4 (1965), by George Lichtheim, reprinted by permission of the Wesleyan University Press,

Quotation, p. 255–256: from *What Coleridge Thought*, by Owen Barfield, © 1971 by Owen Barfield, reprinted by permission of Wesleyan University Press, Middletown, Connecticut.

Quotations, pp. 40, 41, 41–42, 45–47, 48, 50, 51, 52–53, 55–57, 59–60: from *Introduction to Cybernetics*, by W. R. Ashby, Chapman and Hall Ltd., London.

Quotation, p. 281: from *Design for a Brain*, by W.R. Ashby, Chapman and Hall Ltd., London, 1960.

Quotation, p. 131–132: from *Politicians, Bureaucrats, and the Consultant: A Critique of Urban Problem Solving*, by Garry D. Brewer, © 1973 by Basic Books, Inc., Publishers, New York.

Quotations, pp. 7–8, 25, 27, 28, 30, 159–160: from *General System Theory—Foundations Development Applications*, by Ludwig von Bertalanffy, © 1968 by George Braziller, Inc., Publishers, New York.

Quotation, p. 119: from *The Systems Approach*, by C. West Churchman, © 1968 by Dell Books, Publishers, New York.

Quotation, p. 139: from *Economics*, 8th edition, by Paul Samuelson, © 1970 by McGraw-Hill Publishers, New York.

Quotations, pp. 241–242, 243, 244: from *Models of Doom—A Critique of 'The Limits to Growth'*, by H. S. D. Cole, Christopher Freeman, Maue Jahoda, and K. L. R. Pavitt, eds. (New York, Universe Books, 1973), published in Great Britain as *Thinking About The Future* by Chatto & Windrig, Ltd., London. Reprinted by permission.

Quotations, pp. 138, 140, 144–145, 153: from *Main Currents in Modern Economics*, by Ben Seligman (Quadrangle Books, Chicago, 1971).

Contents

The Rise of
Systems Theory

Introduction

A number of disciplines have emerged in the twentieth century that can be classified under the general heading of "systems thinking." These originally separate disciplines include the following:

1. The biological philosophy of Ludwig von Bertalanffy, and his concept of the "open system";
2. Norbert Wiener's formulation of cybernetics and W. Ross Ashby's related work on machines that are claimed to think and to learn and, stemming from this work, the concepts of *feedback* and *automation.*
3. Information and communications theory, based on the work of Shannon, Weaver, Cherry, and others, on the theoretical, mathematical, and linguistic problems involved in the transmission of messages over message-carrying circuits.
4. Operations research, which first emerged full-fledged in England during the War of 1939-1945 under the leadership of E. C. Williams, and has since been institutionalized by the founding of the Operations Research Society of America and the Operational Research Society in Great Britain.
5. The games theory of von Neumann and Morgenstern.
6. The techniques for simulating social and environmental processes by computers, advocated by Jay Forrester and many others.

Technical and scientific advances in different fields have resulted in the establishment of new disciplines. These disciplines have been institutionalized and legitimated by the establishment of scholarly societies, most of which offer yearbooks and have founded journals. Most of these journals offer material of two sharply disparate kinds: (1) articles and studies of a highly technical, narrow, and specific nature, focused on concrete problems within the discipline; (2) large numbers of essays and articles of a "missionary" nature, which are addressed not only to the technicians within the field, but to the general public at large. These articles attempt to explain to the layman the wider significance of the work being done within the discipline. Almost to a man, the practitioners within these fields appear to feel that their work is of more than merely "technical" value. They appear to be convinced that the discoveries and concepts they have developed are of major philosophical, societal, and even religious significance: They offer new images of humanity and society, of God and the creation of human beings, and of their interrelations. In addition to these missionary articles within the professional journals, a third type of publication has emerged—books and anthologies of readings addressed to the public at large, which seek to explain to the lay public what these fields are all about. Through these publications the missionary activities of scientific workers are no longer confined within limited professional media; there is an effort to reach the public at large.

This study is concerned primarily with books and articles of the second and third types, in which systems technicians make philosophical and societal claims on the basis of their technical work. These are examined in detail, especially in terms of their claimed relation to the technical discoveries made in their respective fields. We show also that the language and rhetoric of systems thinking has been adopted by administrators, psychiatrists, social workers, political and social scientists, who themselves have not been engaged in technical systems work, but who adapt these terminologies to their own fields. Such practitioners may perhaps not be called systems men in the narrower technical sense, but may be so designated as a result of their espousal of systems concepts in the broader sense of having general applicability to society at large.

Part I of the study describes the historical emergence of each of these disciplines and their major characteristics. Sociological considerations are secondary to this part; major emphasis is placed on describing the emergence of each discipline and its leading concepts and techniques.

Part II focuses on the societal claims of the systems thinkers. First, we establish that systems disciplines have converged—that is, the leading representatives of these originally separate fields have themselves recognized and proclaimed that they all represent what is basically one unified approach and way of thinking; divergences, they indicate, are based more on the specific features of narrow problems than on intrinsic differences. One indicator of this is the large amount of cross-referencing in the literature as well as the borrowing of terminology; thus, for example, it has become difficult to draw the line between uses of the terms *information theory, communication theory,* and *cybernetics.* Furthermore, many of the leading figures in these disciplines acknowledge as forerunners the same men, and a common lineage in these fields can be traced.

The remainder of Part II focuses on the migration of systems thinking (or at least the terminology thereof) into a variety of administrative and social-science fields, among them organization theory, psychiatry, social work, urban planning, and political theory. The literature of these fields is examined and demonstrates the adoption of systems terminology by practitioners in these fields who are often eager to proclaim that such terminological innovations represent significant progress for their administrative fields.

Part II concludes with a general overview of the emergent view of society seen as a "system" and the claims made by the systems thinkers to have insights into humanity and society that are not granted to others. We suggest that these claims are without substance, that the societal claims of systems thinkers are highly abstract and repetitiously programmatic in the sense of announcing great new insights that they expect to emerge once systems thinking becomes widely adopted. Although some systems thinkers advocate caution in generalizing beyond the limits of their disciplines, they also write "missionary" articles and books for the general reader. The constantly recurring refrain of systems thinkers is that of the great new era that is dawning to replace the present malaise. But what they offer as a view of man in society is not at all new, and it precedes the emergence of their doctrines; for the most part, when the systems thinkers emerge from the discussion of specific technical problems and turn to a philosophy of humanity and society they echo the positivism of Auguste Comte, with a decoration of formal and mathematical terminology.

Part III relates the emergence of systems thinking to the emergence of new scientific and technocratic elites; systems theory is examined from the perspective of Karl Mannheim's concept of ideology; an image of the world that seeks to maximize the social prestige and power of an interest group. The image of the world offered by the systems theorists is shown to be philosophically and scientifically meretricious and unable to stand on its own merits. Its significance, therefore, is not in its intrinsic worth or content, but in the implicit claims to power and prestige contained therein; the material offered has a latent function (i.e., having a purpose separate from the material itself). In Weberian terms the scientific intellectual, like other intellectuals, assumes the mantle of prophecy and priesthood, making claims in a manner similar to the claims that have been made by poets, artists, novelists, journalists, academics, and other intellectuals. Thus, the perpetual claim of the intellectual to unique qualifications as philosopher-king is renewed by the systems scientists, who prefer to be thought of perhaps as scientist-kings.

The man who offers an image of society as a closed system (i.e., able to be encompassed and manipulated by logically closed theoretical models) and who on the basis of technical work and discovery on such systems demonstrates expertise in these matters is clearly offering to assume benevolent control of society as a closed system, which he will manipulate from a position outside of and superior to that system. Such claims have been repeatedly made by scientists on the basis of new knowledge and new discoveries, and since new knowledge and new discoveries are continually emerging, these claims are regularly renewed.

Thus Part III brings ideological and sociological considerations to the fore; a variety of methods are employed to analyze the explicit and implicit meanings of a newly emergent ideology. In addition, historical and cultural considerations demonstrate some of the conditions that permit this ideology to emerge: As earlier validating images of society fade (society as a contract, as a community of minds and values) and as the claims of older political ideologies (Marxism, capitalism, socialism) lose their emotional power, the image of society as a "system" becomes prominent as the newest claimant.

The role of systems thinking and procedures in intensifying the bureaucratization and rigidity of society is clarified; the authoritarian potentialities of systems thinking are made evident, as are the glaring failures and gaps in systems philosophy, which carefully evades the problem of conflicting values in society and the problem of basing freedom on an image of society seen as a system. The only value seen by scientists as intrinsic to scientific research—that is, the freedom of the scientist to investigate scientific problems—does not imply or support the notion of freedom for anyone else. These notions, though proclaimed ritualistically and no doubt sincerely by systems scientists, are not intrinsic to the thought systems they proclaim or to their social possibilities.

The philosophy of systems thinking is for the most part a pastiche of Comtean positivism, of the empiricism of Locke, Berkeley, and Hume in travesty, of modern linguistic philosophy as developed by Wittgenstein, Carnap, Whorf, Quine, and others, overlaid with the assumption that actuarial and statistical logics form the basis for societal processes and that the image of society as a system is a metaphor assumed by systems thinkers and imposed upon the world arbitrarily. Alternative approaches to the conception of humanity and society are suggested.

The gaps and inadequacies of systems thinking as philosophy and as sociology suggest its ideological function, and its relation to the professional, bureaucratic, and budgetary interests of systems writers indicates its significance for society at large.

I am indebted to many for their help at one or another stage of work: to Joseph Bensman, Ida Hoos, Stanford Lyman, Emil Oestereicher, Zoltan Tar, and Arthur Vidich, for many perceptive and constructive comments; and to Estelle Cooper for careful typing and editing.

The Disciplinary Origins of Systems Theory

The Emergence of Systems Theory

S ystems theory claims to be a major reorientation in scientific thought, of the type Thomas S. Kuhn describes as a "scientific revolution."[1] Kuhn describes the history of science as consisting of tradition-bound periods that are punctuated by changes of a noncumulative nature. Thus, within a given period, the overall framework or point of view of the field remains fixed and stable; scientific work usually consists of applying and elaborating concepts that are taken for granted. New work is done and new discoveries occur without overthrowing the general framework or point of view. But at some point discoveries occur, the implications of which are "revolutionary" in that they suggest the overthrow of the paradigm or general conceptual framework within which scientific work has been done.

A period of chaos then ensues, while the scientists do their best to "pick up the pieces" and to create a new conceptual framework, or paradigm. Once this is done, scientific work of a more "routine" nature resumes; much of the scientific work is more of the nature of filling in details, and measuring phenomena more accurately within what is becoming the next tradition-bound period.

Systems theorists claim that their work constitutes just such an overthrow of the hitherto accepted framework of science; scientific thinking must be reoriented accordingly.

The nature of the revolution is the replacement of analytic atomistic modes of thought by holistic integrative ones. One of the founders of systems theory, Ludwig von Bertalanffy, describes it in the following terms:

The 19th and first half of the 20th century conceived of the *world as chaos*. Chaos was the oft-quoted blind play of atoms, which, in mechanistic and positivistic philosophy, appeared to represent ultimate reality, with life as an accidental product of physical processes, and mind as an epiphenomenon. It was chaos when, in the current theory of evolution, the living world appeared as a product of chance, the outcome of random mutations and survival in the mill of natural selection. In the same sense, human personality, in the theories of behaviorism as well as of psychoanalysis, was considered a chance product of nature and nurture, of a mixture of genes and an accidental sequence of events from early childhood to maturity.

Now we are looking for another basic outlook on the world—*the world as organization*. Such a conception—if it can be substantiated—would indeed change the basic categories upon which scientific thought rests, and profoundly influence practical attitudes.

This trend is marked by the emergence of a bundle of new disciplines such as cybernetics, information theory, general system theory, theories of games, of decisions, of queuing and others; in practical application, systems analysis, systems engineering, operations research, etc. They are different in basic assumptions, mathematical techniques and aims, and they are often unsatisfactory and sometimes contradictory. They agree, however, in being concerned, in one way or the other, with "systems," "wholes" or "organization;" and in their totality, they herald a new approach.[2]

The emergence of systems theory appears to have proceeded in stages; first, there were a number of anticipations by philosophers and psychologists; then the full-fledged statements by von Bertalanffy that established systems theory as a movement in biology and physics. We examine the emergence of systems theory in its main outlines by tracing the principal points made in its major documents.

ANTICIPATIONS OF SYSTEMS THEORY

THE CONTEXTUALIST AND ORGANICIST APPROACHES OF STEPHEN C. PEPPER

In his *World Hypotheses* Pepper attempted to describe the principal metaphysical systems in their efforts to comprehend and account for the world of experience.[3] Pepper describes six major "root metaphors," of which only four, he maintains, are worthy of being taken seriously; He discredits the first two, mysticism and dogmatism. As a source of knowledge mysticism is simply too private to hold for those who do not share mystical experiences. Dogmatism is a bit more complicated. Dogmatists often rely on "infallible authorities"; but infallible authorities often contradict one another, or they may base their authority on an appeal to self-evidence, to commonsense principles, which on closer examination no longer appear self-evident, but rather are seen as arbitrary constructions that by custom and usage are taken for granted. (By way of illustration, Pepper offers the assumptions of Euclidean geometry.) Dogmatism, for Pepper, is easily discredited by epistemological considerations.[4]

This is not to say that dogmatic social authorities have no legitimate place in society; we trust them because we regard them as reliable transmitters of information, but the validity of the information is determined by other criteria.

Having established an epistemology closely related to scientific method in his third chapter and disposing of animism along the way, Pepper develops a metaphysics of world images, or world hypotheses. For Pepper, only four world hypotheses are relatively adequate; each one is determined by a specific root metaphor, each is autonomous and is able to supply a conceptual framework adequate for its data without borrowing from the others, each is sufficiently wide in scope at the same time that it offers a satisfactory level of precision.

The first world image is *formism*, familiar to us as Platonic realism. Objects of experience are seen as copies of ideal forms, and a total world view can be built up along the lines of such essences or categories.

The second view is *mechanism*, which we know primarily in terms of the Newtonian world view, in which material particles operating under physical laws establish a mechanist and wholly determined world.

The third and fourth metaphysical systems, contextualism and organicism, are more closely related to the emergent philosophy of systems thinking.

Contextualism and Organicism

Contextualism is the name given by Pepper to the metaphysics inherent in the early pragmatism of Peirce and James. The world is seen as an unlimited complex of change and novelty, order and disorder. Out of this total flux we select certain contexts; these contexts serve as organizing gestalts or patterns that give meaning and scope to a vast array of details that, without the organizing pattern, would be meaningless or invisible. Thus, an organizing context creates a "theme" that has texture, quality, detail, and a "specious present." Furthermore, it "fuses" into a unity items that in other contexts appear as discrete entities. Thus, in focusing upon the meaning of a sentence, we tend to ignore the specific words of the sentence, unless we deliberately select such a "strand" in the given "texture" to analyze.

Important here are the concepts of quality and fusion. A given quality

always exhibits some degree of fusion of the details of its texture. This feature is perhaps most clearly perceived in savors and musical chords. William James' lemonade has become famous in this regard: lemon, sugar, and water are the ingredients or details of the taste, but the quality of lemonade is such a persistent fusion of these that it is difficult to analyze the components. A simple musical chord is perhaps a still better illustration because most people can voluntarily take it as either fused or unfused. The tonic triad C-E-G has a distinctive character. Most of us hear it strongly fused and recognize it at once for its distinctive quality, just as we recognize lemonade. Flat the E, and another chord is felt which has another highly distinctive quality. But with a shift of attitude the C-E-G chord can be perceived relatively unfused . . . The event quality in the two perceptions is quite different. Where fusion occurs, the qualities of the details are completely merged in the quality of the whole. Where fusion is released, the details take on qualities of their own, which may in turn be fusions of details lying within these latter qualities. Fusion, in other words, is an agency of qualitative simplification and organization . . . some fusion must remain in the quality of an event; otherwise the event would break apart and we would have not a single event, but two quite unconnected events . . . Contextualism is the only theory that takes fusion seriously. In other theories it is interpreted away as vagueness, confusion, failure to discriminate, muddleness.[5]

Thus, in this theory, we "organize" our experience by adopting "themes" or contexts; within these contexts meanings emerge in complex strands or levels that would disappear without the organizing context. When strands converge or diverge, we speak of "similarities" or of "contrasts."

From the assumptions of contextualism a specific theory of truth emerges— operationalism, an outgrowth of the pragmatism of James and Dewey. Truth is "the successful working of an idea"[6] within a specific (and always limited) context. Truth is verification in practice. The contextualist hesitates to extend a theory beyond the limits of specific "working" contexts; all experiences are fragmentary, limited, partial, and occur within the limits of contexts beyond which only an infinite universe of indeterminancy. No theories or formulations drawn from within a limited context (and all contexts are limited; none encompass the world) can be used to construct a theory or a metaphysics that will successfully digest the world. All conceptual schemes occur within a universe and can never grasp the total structure of events. Many contextualists would even deny that the universe has a "structure" that can be grasped. Nature itself may be constantly changing and full of novelties.

The philosophical and epistemological problems connected with this world view are beyond the scope of this study; what is relevant to the development of systems theory is the "organizing" properties of contextualism; The world of experience is seen as a chaos of potentialities that spring into a meaningful form only under an organizing context. Parts are meaningless when detached from a whole; more than that, they are not only meaningless but often simply unperceived or unperceivable.

The similarity of this view to what is known as gestalt psychology comes immediately to mind; in fact, some writers explicitly regard gestalt psychology as a major forerunner of systems philosophy.[7]

Contextualism, important as it is, is not the only root metaphor that relates to systems theory.

Organicism

The fourth root metaphor offered by Pepper, organicism, also constitutes a major orientation for some systems theorists.

The contextualist uses the category of integrating structures (contexts) to explain experience, but denies to these integrating structures any reality of significance. The organicist maintains that "integrating structures surrounding and extending through given events"[8] are more numerous, coherent, and "real" than the contextualist wants to admit. Our experience is not the chaos the contextualist would have us believe, but shows undeniable regularities of detail and texture. For the contextualist, the truth of any idea or theory is operational; these ideas are never firmly established, but may at any time be overthrown by the emergence or discovery of a new fact. The chagrin of the philosopher-scientist at the overthrow of a theory by a new discovery results, in the view of the contextualist, only because the philosopher has taken the concepts too seriously.

The organicist, by way of rebuttal, points to the same scientific theories to show that in fact the overthrow of the scientific theory does not mean a collapse into chaos, but rather the replacement of a relatively limited integrating form by a more comprehensive and more accurate form. The overthrow of the Ptolemaic theory by the Copernican theory serves as a warning to the contextualist of the fragility of all theories; it serves as a sign to the organicist that larger and more comprehensive integrating forms are always being discovered. Thus, the materials of experience are never lost when one scientific world view is overthrown by another; rather, they are, in Pepper's terms, transferred from a system in which they did not belong to one in which they do belong.[9]

Each level of integration resolves the contradictions of the levels below and so removes the errors that were most serious there. Each level brings about an improvement of judgment. Each level exhibits more truth through higher integration of the facts. There is much more truth in Ptolemy than in Anaximenes, more in Kepler than in Ptolemy, more in Newton than in Kepler. It appears that the criteria of truth are precisely the features of the organic whole—inclusiveness, determination, and organicity . . . this theory of truth is known as the *coherence theory*. It is obviously implied by the categories of organicism and obviously presupposes those categories. In other views coherence may be treated as a gauge of truth but not as its essential nature. . . .[10]

More important, perhaps, the coherence of organicist theory is not the formal self-consistency of logic, but the "organic relatedness of material facts."[11]

Pepper describes the limitations of the organicist position as arising in certain contrasts with contextualism: No matter how much "progress" is made in developing ever more comprehensive insights or theories, the experience of incompleteness and contradiction remains a constant. Furthermore, the organicist consistently underrates the phenomena of uniqueness and of historicity. He tends to ignore time. By way of rebuttal, he contends that the contextualist position, in its overemphasis on temporality, uniqueness, and concreteness, ignores the integrations demonstratable in the history of science and philosophy.

□　□　□

LAWRENCE J. HENDERSON

Henderson (1878–1942) may also be mentioned as a "forerunner" of systems thinking; his early career was in biochemistry; he was graduated from Harvard Medical School in 1902; did further studies in chemistry in Europe, and returned to Harvard to teach biochemistry. He did distinguished work in the field and later in life turned to the philosophy of science and finally to the teaching of sociology. His sociological writings were not numerous, or widely published in his lifetime; but through his lectures at Harvard, especially the *Sociology 23* lectures, and through his association with the Society of Fellows, his influence became important for a number of distinguished and highly influential social scientists and scholars, including Talcott Parsons, Elton Mayo, T. North Whitehead, Fritz J. Roethlisberger, Chester I. Barnard, George C. Homans, William F. Whyte, Conrad M. Arnesburg, Crane Brinton, and Robert K. Merton.[12]

Henderson based his sociological thinking partly on biochemical and physiological analogies, partly on the concept of system expressed by the American physicist Josiah Willard Gibbs, and partly on the sociology of Pareto; in fact he described his encounter with Pareto's ideas as a major event in his intellectual life.

His scientific philosophy is strongly pragmatic in character; in Pepper's terms he is a contextualist. For Henderson, theoretical concepts used in science are constructs having only and always a provisional value. Associated with this is his antirationalism: Men put too much faith in the reasonableness of their ideas and actions and underestimate the pervasiveness of irrational (or nonrational) sentiments in their thoughts and actions. Much of Henderson's writing appears in this sense to be similar in tone to those of the logical positivists in that he regards ethical judgments, along with a large variety of classes of statements and sentiments, as meaningless in the sense that no operational procedure can be developed by which they can be tested.[13]

Though he stressed the power of nonrational thought and action in human life (under the impact of Pareto's thinking), Henderson in effect acted to devalue the nonrational by stressing its operational meaninglessness. In this respect his writings give no impression of originality; they breathe an air of naive scientism which made his political comments especially superficial. Thus, he appears to have praised Mussolini as a practitioner of his much-admired Pareto's point of view.[14]

What gives Henderson a place in the history of systems theory is his insistence on regarding social processes in systems terms; here he claims indebtedness especially to the French physiologist Claude Bernard as well as to Gibbs and to Pareto. Henderson wrote an introduction to the 1949 reprint of Bernard's *An Introduction to the Study of Experimental Medicine.*[15]

The concept stressed by Henderson is that of equilibrium; for him, this concept, stemming equally from the work of Gibbs, Bernard, and Pareto, was essential to the study and understanding of social processes.

Equilibrium

The organism possesses a self-regulating mechanism whose goal is the maintenance of equilibrium (health); a condition of disequilibrium defines illness. Henderson's definition of equilibrium is borrowed from Pareto and importantly related to later systems concepts.

A state such that if a small modification different from that which will otherwise occur is impressed upon a system, a reaction will at once appear tending toward the conditions that would have existed if the modification had not been impressed. . . . Equilibrium is an equilibrium of forces, more or less like the equilibrium, for instance, in a box spring; that a small modification leaves the forces substantially intact; and that the forces tend to re-establish the state that would have existed if no modification had occurred.[16]

The penchant of systems thinkers for arguing from geometrical diagrams and from simple mathematical formulas is foreshadowed in Henderson's work.

In a social system all factors (persons, interests, residues, etc.) are mutually dependent or interactive. In order to fix our ideas, let us consider a relatively simple mechanical system. It may seem that we are reasoning from analogy, but this is not so. On the contrary, we shall be reasoning logically from premises stated above, because the mathematical formulation necessary to describe this mechanical system would be formally identical with that necessary to describe the analogous social system.

In Figure 1 let *A, B,* and *C* be objects connected together and to the rigid framework *A, B, C* by elastic bands 1, 2, 3, 4, 5, 6. Further, let each band be attached in *A, B,* and *C* to some kind of mechanism which constantly operates in some determinate manner to wind up or to unwind the elastic band, and let this action be a function of time and also of the instantaneous tensions. In other words, let each winding mechanism run on in some manner predetermined so that the winding shall in each instance vary quite definitely with time and with tension, and with nothing else directly. Further, let the masses of *A, B,* and *C* be known.[17]

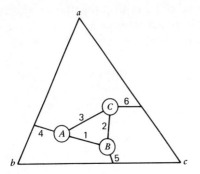

Henderson continues to the effect that such a system can, at least in principle, be described mathematically and that such a description makes deduction of the state of the system possible at any instant. Everything in the system is dependent on the previous state of the system. Going further, however, he explicitly states that we can consider *A, B,* and *C* to be individuals in a social system, and the bands, their tensions, and mechanisms can be replaced by properties of social systems. If a displacement is introduced into the system, various motions will occur, but the system will tend to return to its original state; this he calls "equilibrium" and explicitly argues that this is a way of conceiving of the social system. All factors interact in a social system; everything is in a state of mutual dependency on everything else.

Elsewhere, in his *Sociology 23* lectures, Henderson made clear his feeling that Pareto's social system was "analogous in its usefulness to Gibbs' physico-chemical system."[18]

However, Henderson's embrace of Pareto's thought did not give him any deeper insights into politics.

I hope it will now be clear that the prevalent description of Pareto as the Karl Marx of the bourgeoisie or of fascism is nothing more than a derivation. It is a fact that Signor Mussolini has attributed his abandonment of socialism to the teaching of Pareto. It is also true that among the Fascists and the Nazis, Pareto's work is much esteemed, though perhaps not always understood. But his writings are no less applicable to France, England, the United States, and Russia than to Italy and Germany, and Pareto himself preferred to all other governments those of some of the smaller Swiss cantons.[19]

Many characteristics of systems thinkers are foreshadowed in Henderson's work, in addition to his early and influential use of the term *system:* his scientism, his passion for quantification, and his enthusiastic and somewhat simplistic belief that systems models can adequately encompass the totality of a society.

WALTER B. CANNON

Henderson's friend and colleague at Harvard, Walter B. Cannon, is also claimed by Henderson and others to be an important forerunner of systems thinking. Most of his writings recorded his medical researches as a physician and professor of medicine at Harvard Medical School. But in his widely praised *The Wisdom of the Body*, Cannon developed ideas that were in fact influential for the development of systems thinking.[20]

The main theme of Cannon's book is the concept of *homeostasis:* A variety of mechanisms exist in the organism to maintain fixed levels of blood sugar, blood proteins, fat, and calcium, as well as adequate supplies of oxygen, a constant body temperature; homeostasis is a condition that may vary, but is relatively constant. Many such conditions may deviate from "normal" levels under, for example, sudden exertion, extreme heat or cold, the occurrence of a wound, the perception of danger, or the experimental removal of organs. Most of Cannon's book is devoted to a description of portions of the organism—nervous system, circulatory system, endocrine glands among them—that operate to produce and restore homeostasis; in addition, he describes the experimental bases for these findings. Thus far his book remains within the realm of physiology. But in his conclusion ("Epilogue—Relations of Biological and Social Homeostasis") Cannon argues from the biological to the social.

Are there not general principles of stabilization? May not the devices developed in the organism for preserving steady states illustrate methods which are used, or could be used, elsewhere? Would not a comparative study of stabilizing processes be suggestive? Might it not be useful to examine other forms of organization—industrial, domestic or social—in the light of the organization of the body?[21]

Insights derived from physiology might prove fruitful for the study of society. His analogy develops into an argument against individualism: Although individual cells engage in a certain amount of self-regulation, in the more complex organisms these cells become fixed in place in specific organs and in effect resign the problems of "getting food, water and oxygen, avoiding too great heat or cold, and keeping clear of the dangers of accumulating waste"[22] to the central nervous system. The analogy is drawn, in sociological terms, with primitive food-gatherers who move about and forage for themselves, but are dependent upon their immediate environment for sustenance. Only with the development of more complex social systems and a more complex division of labor do the individuals become more mutually dependent, gaining thereby the advantage of relative freedom from the immediate pressures of food-gathering. Therefore, just as with "the body physiologic," the "body politic" benefits by the development of central controlling mechanisms.

He draws a direct analogy between the "fluid matrix" of animal organisms and the transportation system of a state or nation—railroads, canals, rivers, roads—with the boats, trucks, and trains serving as common carriers, like the blood and the lymph; wholesalers and retailers represent the less mobile portions of the system. Thus, products of farm, factory, and the like are carried back and forth. Money and credit, insofar as they facilitate these exchanges, are integral parts of the fluid matrix of a society.[23]

To stabilize the social organism, Cannon argues, this fluid matrix should ensure the flow of necessities to all members of the social organism. As excessive toxins may accumulate in the organism, there may be overproduction of goods in the economy, and dispersal presents a problem. The central administrative organs of the body politic must deal with this problem. The analogy is developed further: As "medicosocial" intelligence has eliminated plagues, malaria, and so on, it may solve other problems. These achievements involve a lessening of the independence of individual members in favor of social control and organization. But the advantages are evident: Persons in need due to hunger, fear, insecurity, are not free. Banishment of such distress would mean their liberation. "The assurance of freedom *to men who are willing to work* would justify a larger control of economic processes, repugnant though that may seem, for it would be a sacrifice of lesser for greater values."[24]

Fears of execssive monotony are unjustified; freedom from concern with daily necessities would give men and women a higher freedom, serenity, and leisure, the primary conditions for wholesome recreation and for the cultivation of individual aptitudes.[25]

LUDWIG VON BERTALANFFY

The work of men like Henderson, Cannon, Koehler, and Angyal, important though it is for establishing the foundations of system theory, is generally considered by systems theorists to be prefatory work. The formulations of the concept of an open system by Ludwig von Bertalanffy (1901–1972) first established systems thinking as a major scientific movement. Similar to Henderson and Cannon, von Bertalanffy began his career as a scientist, making contributions to theoretical biology from the 1920s on; later his philosophical interests (which in fact also emerge relatively early) began to overshadow his work in biology *per se*.[26]

His first formulations of system theory emerge full-blown in relatively early works such as his *Modern Theories of Development* (1933) and in his *Problems of Life* (1952).[27] But his essay "The Theory of Open Systems in Physics and Biology" (printed in *Science*) (1950) gave rise, in Emery's words, to "the movement of thought that has sustained the *General Systems Yearbook* for more than a decade."[28]

In this formulation von Bertalanffy sought to establish the foundations for systems thinking on a biological, rather than a philosophical or merely formalistic basis.[29]

Von Bertalanffy has long maintained that the phenomena of biology call for new ways of thinking, that the methods of the physical sciences are inappropriate. Thus in *Modern Theories of Development* he argues that organic laws, in contrast with physical ones, require a new kind of statistics, of "higher order," which will not be resolvable into the assumptions of statistics.[30]

Furthermore, he maintains that the fundamental problem for modern biology is the discovery of "the laws of biological systems to which the ingredient parts and processes are subordinate."[31] Thus, the present antithesis between mechanism and vitalism can be overcome. Much of *Modern Theories of Development* was given over to a discussion of work in experimental biology designed to show that biological organisms or cells, when interfered with experimentally, could correct for the interferences and arrive at fully developed animals; thus, transplanting regenerative buds from a newt's tail to the neighborhood of a leg produced not a tail but a leg.[32] Nothing in the laws of physics could account for such phenomena. Von Bertalanffy concluded that higher levels of organization involve new laws that are not deducible from the laws appropriate to the lower levels. The hierarchical mode of organization must have greater significance for living organisms than for inorganic objects and must, therefore, be fundamental to biological law.[33]

He offered a similar theme in *Problems of Life*. Science is seen as a hierarchy of statistics.

All laws of nature are of a statistical nature. They are statements about the average behavior of collectives. Science as a whole appears as a hierarchy of statistics.

At the first level is the statistics of microphysics. . . . A second level is constituted by the laws of macrophysics. . . . A still higher level is represented by the biological realm. . . . Finally there are the laws that apply to the supra-individual units of life . . . laws of this kind are the basis for insurance statistics, and hence are of great practical and commercial importance.[34]

Von Bertalanffy makes clear, as he does in most of his writings, his belief in the fundamental unity of the sciences. Physics, biology, psychology, philosophy are all to culminate in a general system theory. In this connection he cites phenomena of an interdisciplinary nature—for example: "relaxation oscillations appear in certain physical systems, and in many biological and demographical phenomena as well."[35] Thus, a general systems theory is a step toward the *Mathesis Universalis* of which Leibniz dreamed.

The parallels he finds in these various spheres are more than mere analogies; we must distinguish between three levels:

1. *Analogies.* These are scientifically worthless—superficial similarities in phenomena that do not correspond to underlying factors or laws operating in them.

2. *Logical homologies.* Here phenomena differ in the causal factors involved, but are governed by structurally identical laws—for example, the flow of fluids and of heat conduction are expressed by the same law.

3. *Explanation* in the proper sense, dealing with the appropriate conditions and laws.

General systems theory, according to von Bertalanffy, serves as a tool for distinguishing analogies from homologies; it leads to the transfer of laws from one field to another, while screening out incorrect analogies. How this is to be done is not specified.[36]

THE OPEN SYSTEM

Von Bertalanffy's essay "The Theory of Open Systems in Physics and Biology" established systems theory as a scientific movement.[37] The main concepts are as follows:

1. "The characteristic state of the living organism is that of an open system." It is open in the sense that it exchanges material with its environment; by this import and export of materials, there is change of components. Previous conceptions of the organism as maintaining a state of equilibrium must yield to the idea of the *steady state.*

2. The concept of the open system maintaining itself in a steady state represents a departure from the concepts of classical physics, which has dealt for the most part with closed systems. According to the second law of thermodynamics, a closed system must eventually attain a state of equilibrium with maximum entropy and minimum free energy. But under certain conditions an open system may maintain itself under a steady state.

3. The mathematics of the steady state may be developed based upon the nature of the chemical reactions going on within the system, some of which may be reversible.

These characteristics of steady states are exactly those of organic metabolism. . . . There is first maintenance of a constant ratio of the components in a continuous flow of materials. Second, the composition is independent of, and maintained constant in, a varying import of materials; this corresponds to the fact that even in varying nutrition and at different absolute sizes the composition of the organism remains constant. Third, after a disturbance, a stimulus, the system re-establishes its steady state. Thus, the basic characteristics of self-regulation are general properties of open system.[38]

The mathematics of the import of materials into the system, and their use for the creation of specific compounds—some of which are retained in the system, others of which are transported outside the system—indicates (a) that the composition of the system in the steady state remains constant even though some of the reactions are irreversible; (b) that the steady state ratio of the components depends only on the system constants and not on the environmental conditions, and (c) that the system may, in the presence of a "stimulus" from outside, a disturbance, manifest forces that counteract the disturbance of the steady state. To maintain itself in its steady state the system requires a constant supply of energy.

4. A profound difference between most inanimate, or closed, systems and living systems is expressed by the concept of "equifinality." In an inanimate system the final state of the system is determined by its initial conditions. A change in the initial conditions produces a change in the final conditions. A different behavior is shown among vital phenomena: Under many conditions the same final state may be reached from different initial conditions and in different ways. Thus, in its early stages the embryo of a sea urchin may be altered by the transplantation of some of its cells; the result will be a sea urchin indistinguishable from one whose embryo has not been so altered. Though equifinality is not a proof of vitalism, it can be shown that equifinality is not to be found in closed systems, which is why it is not generally found in inanimate systems. The equations of steady state systems show that the initial conditions do not appear in the steady state, whose values are always the same, "being determined only by the constants of the reactions and of the inflow and outflow." In some biological cases equifinality can be formulated quantitatively.

Growth can be considered the result of a counteraction of the anabolism and catabolism of building materials. In the most common type of growth anabolism is a function of surface, catabolism of body mass. With increasing size the surface-volume ratio is shifted in disfavor of the surface. Therefore, eventually a balance between anabolism and catabolism is reached which is independent of the initial size and depends only on the species-specific ratio of the metabolic constants. It is therefore, final.

Study of the open system has an important bearing on thermodynamics, especially with reference to entropy. Closed systems, characterized by irreversible processes, tend toward increased "positive" entropy (the loss of energy). But in open systems, especially in the living organism, both "positive" and "negative" entropy occurs; the organism feeds upon this negative entropy by importing complex organic molecules, using their energy and returning the simpler products back to the environment. Thus, the second law of dynamics, which applies to the universe as a whole and to the open system plus its environment, need not apply to the open system itself. The second law of thermodynamics may be stated in another way: The general trend of the universe is toward states of maximum disorder and the leveling of differences; the higher forms of energy, such as mechanical, chemical, and heat gradients, will continually disappear. Thus, the universe will approach entropy death when all energy is converted into heat of low temperature, and the world process will come to an end. Though exceptions to the second law exist in some microphysical dimensions, such as the interior of stars, the general trend degradation of energy appears to be a necessary consequence of the second law.

At this point the contrast between inanimate and animate nature becomes evident.

In organic development and evolution, a transition toward states of higher order and differentiation seems to occur. The tendency toward increasing complication has been indicated as a primary characteristic of the living, as opposed to inanimate, nature. . . . These problems acquire new aspects if we pass from closed systems, solely taken into account by classical thermodynamics, to open systems. Entropy may decrease in open systems. Therefore, such systems may spontaneously develop towards states of greater heterogeneity and complexity. . . . Probably it is just the thermodynamical characteristic of organisms as open systems that is at the basis of the apparent contrast of catamorphosis in inanimate, and anamorphosis in living, nature. This is obviously so for the transition toward higher complexity in development, which is possible only at the expense of energies won by oxidation and other energy-yielding processes.

Thus, the theories of macrophysics must be complemented by the thermodynamics of open systems. "Not only must biological theory be based upon physics; the new developments show that the biological point of view opens new pathways in physical theory as well."

Von Bertalanffy then indicates the applications of the theory of open systems to biology. Organisms are "quasi-stationary open systems." Phenomena such as metabolism, irritability, and autonomous activities may be understood as maintenance of the steady state, while "growth, development, senescence and death represent the approach to, and slow changes of, the steady state." He then surveys briefly work going on in experimental and theoretical biology relevant to systems theory; these include a theoretical cell model, into which substances flow from outside and undergo chemical reactions and reaction products flow out. Consequences are derived from this model that correspond to the basic characteristics of the living cell, including growth, periodic division, order of magnitude similar to average cell size, and the possibility of nonspherical shapes. Other workers in the field focus upon the conditions by which the cell can generate internal conditions different from that of the surrounding medium; they selectively accumulate salts, and can change size. All of these are conditions of an open system attaining a steady state. Still others have worked on the mathematics of the open system, and on chromosomal and genetic aspects. Tracer studies of metabolism especially have helped popularize the concept of the organism as a steady state. Chemists develop "gross formulas" that indicate the net results of long chains of complex and "partly unknown reactions."[39]

Thus, even where we do not know the specifics of functioning in detail, by overall statistical methods we may be able to comprehend the system as a whole. The affinity of this type of methodology and conceptualization to managerial and actuarial approaches to society is, of course, the theme of this study. Other areas in which fruitful results have been obtained include the quantitative theory of growth, the steady state and turnover rate of tissues, and studies in excitation seen as a reversible disturbance of processes going on in the organism.

Norbert Wiener's much discussed work in cybernetics, the theory of feedback mechanisms, is related to the theory of open systems. Feedbacks, both in manmade machines and in organisms, are based upon structural arrangements. Such mechanisms in the organism are responsible for homestasis. Other areas in which systems thinking is fruitful include pharmacology, demography, and sociology. Von Bertalanffy concludes with a statement that a general systems theory is the product of the formal correspondence of general principles observed in various fields. In physics the theory of open systems will open new avenues; in biology, it is at the basis of life phenomena.

This essay by von Bertalanffy, published in 1952, is said to have focused considerable interest in the scientific community on the concept of general systems. In 1954 von Bertalanffy and scholars such as Anatol Rapoport and Kenneth Boulding formed a Society for General System Theory; the society later changed its name to the Society for General Systems Research. Its annual yearbooks, published since 1956, are an important source of documents related to systems thinking. Von Bertalanffy's work has never been purely that of a biologist; it has included philosophical, psychological, and sociocultural themes from the very beginning. Since our concern with von Bertalanffy's work is not primarily with biology, but with the claims of systems thinkers with respect to society, we must also examine the ways in which von Bertalanffy has extended these claims beyond the realm of biology; the philosophical and sociological aspects of his work must be examined along with the biological. Von Bertalanffy has repeatedly (and somewhat repetitiously) summarized the claims of systems theory in a number of works subsequent to the essay summarized above. This include his *Robots, Men, and Minds* (1967), addressed to the problems of psychology; a number of essays in the *General Systems Yearbook;* and especially his *General Systems Theory Foundations Development Applications* (1968). This latter in many respects pulls together and summarizes his world view, though without adding anything new with respect to biology.

To the system properties described in the above essay, von Bertalanffy has added a number of more formal, and even metaphysical, concepts concerning the properties of a system. In his essay of 1962 ("General Systems Theory—a Critical Review")[40] von Bertalanffy extends and systematizes his philosophy, as he does in his book *General System Theory.* His philosophical and societal claims on behalf of systems theory show little change over time, and so may be easily summarized.

The Inadequacy of Classical Physics to Account for Biological, Psychological, and Social Phenomena

Both in living organisms and in human behavior, we see order, regularization, self-maintenance during continual change, regulation, and apparent teleology. In human behavior we see goal seeking and purposiveness. The urgent scientific question is whether conceptual schemes can be expanded to deal with these problems where the application of physics proves insufficient or unfeasible.

The mathematics of classical physics proves inadequate for biological and societal phenomena. Classical mechanics was concerned with simple one-way causal relationships and with relationships between few variables. It could give perfect solutions for the attraction between two celestial bodies and hence permits the exact prediction of future constellations or even the existence of hitherto undetected planets. But the three-body problem is unsolvable in principle and yields only to approximations. Similarly in modern atomic physics two-body problems such as the relation of proton to electron are solvable, but trouble arises with the many-body problem. Because many problems in the biological, behavioral, and social sciences are essentially multivariate problems, new conceptual tools are needed. The mathematics of classical physics is essentially geared for unorganized complexity, but in modern physics and biology the problems concern the interaction of a large but not infinite number of variables which demand new conceptual tools.

The essential unsolvability of complex mathematical problems by analytical methods can be transcended by the advent of the computer. Systems far exceeding the reach of conventional mathematics can be computerized, and actual laboratory experiments can be replaced by computer simulation. In such a way, for example, has been calculated "the fourteen-step chain reaction of glycolysis in the cell in a model of more than 100 nonlinear differential equations. Similar analyses are routine in economics, market research, etc."[41]

Thus an expansion of science is necessary to deal with the aspects that are left out in classical physics and conern only the specific characteristc of biological, behavioral, and social phenomena. New conceptual models must be introduced.

Interdisciplinary Theories Representing an Advance beyond Classical Physics

When von Bertalanffy began work in science, biology was embroiled in the controversy between mechanism and vitalism. In attempting to resolve this controversy he first formulated organismic and system concepts. Because the climate of opinion was unfavorable, he left his materials unpublished until after the war. By then the intellectual climate had changed, making abstract model building and generalizations more fashionable. Thus, novel and independent developments had occurred, supporting von Bertalanffy's original insights.[42] These fields include cybernetics, information theory, game theory, decision theory, topology or relational mathematics, factor analysis, systems engineering, operations research, social work, and human engineering. In addition, there is general system theory in a narrower sense (GST), which tries to derive from a general definition a "system" as a complex of interacting components concepts characteristic of organized wholes such as interaction, sum, mechanization, centralization, competition, finality, and so on and to apply them to concrete phenomena. Such concepts, which appeared in mechanistic science to be metaphysical and unscientific, are today taken · seriously and seen as amenable to scientific analysis.

General System Theory In the Broad Sense the Basic Science of Which Systems Engineering, Operations Research, and Human Engineering Represent Applied Science

There is a general tendency towards integration in the various sciences; this integration appears to be centered in a general theory of systems; it may be an important means for aiming at exact theory in the nonphysical fields of science and brings us nearer to the goal of the unity of science by developing unifying principles that run "vertically" through the universe of the individual sciences; this can lead to a much-needed integration in scientific education.

System properties can be stated in a set of mathematical forms, and these forms constitute a set of isomorphisms—that is, they are found to operate and to have applications across a variety of fields.

Thus, a system can be defined mathematically by a system of simultaneous differential equations such that a change of any one measure within the system is a function of all the other measures within the system; conversely, change of any one measure entails change of all the other measures and of the system as a whole.

Minor alterations of the system of equations can result in specifying the constants for the steady state of the system, assuming that the system can be developed in Taylor series. This assumption also makes it possible to develop a general solution for the system of equations and to specify the roots of the equation.[44] Depending upon the mathematical properties of the roots (real or imaginary, positive, zero, or negative), the system can be characterized as periodic, cyclical, nodal, and so on.

Von Bertalanffy thus arrives at a series of simple curves, among them exponential growth curves. He then repeats and extends his claim that these system models apply "in a variety of fields, and we can use (the system of equations developed) to illustrate the formal identity of system laws in various realms, in other words, to demonstrate the existence of a general system theory."[45]

In mathematics the exponential law is called the law of natural growth and is valid for the growth of capital by compound interest. He finds applications for these laws in biology, sociology, chemistry, demography, and ecology.[46]

Von Bertalanffy's concept of sociology and of social science in general, as revealed here and in the material to follow, would not be recognizable to many practitioners in these fields.

By other manipulations of these "mathematically trivial" equations[47] he also arrives at the logistic curve, also of wide application. In chemistry it is the curve of autocatalytic reaction (in which the reaction product obtained accelerates its own production). "In sociology, it is the law of Verhulst . . . describing the growth of human populations with limited resources." Though these examples may be mathematically trivial, they illustrate that laws of nature can be arrived at "not only on the basis of experience" but by interpretation. This is claimed to show the existence of a general system theory.[48]

A rudimentary set of equations is thus used as a basis for claims for insight into the nature of the universe and claims for the ability to unify separate spheres of being, knowledge, and thought.

By similar operations von Bertalanffy proceeds to show that a set of somewhat simplistic mathematical models can be created that describe growth rates of parts of an organism; the competition for limited resources both within an organism and within a nation (he cites Pareto's law on the distribution of income within a nation); the conditions under which systems display properties such as wholeness (the extent to which parts of a system are dependent upon other parts of the system), centralization (the extent to which one component of a system dominates all other components), and summativity (the extent to which change in the total system obeys an equation of the same form as the equations for the parts).

Thus von Bertalanffy attempts to develop a set of theoretical concepts based upon a simplified mathematics of systems, based also on the assumption of their applicability in various spheres of experience, which he claims can culminate in a unification of the sciences. Central to his thinking is the belief in isomorphisms:[49] The same laws find expression in different and apparently unrelated fields. He again cites the same examples: The principle in mechanics known as minimum action appears in physical chemistry as Le Chatelier's principle, in electricity as Lenz's rule, in population theory according to Volterra, and so on. General system theory, then, will serve as an "important regulative device in science." It will make possible the transfer of simplified conceptual models from one field to another, "and it will no longer be necessary to duplicate or triplicate the discovery of the same principle in different fields isolated from each other. At the same time, by formulating exact criteria, general system theory will guard against superficial analogies which are useless in science and harmful in their practical consequences."[50]

The development of isomorphisms in separate fields is understandable with reference to linguistics and to evolution. The development of primitive languages among mutually isolated groups shows, he says, sound mutations that are strikingly similar, as are independent evolution of groups within given classes of mammals.

General system theory, then, will be a discipline that develops, tests, and demonstrates laws that apply equally to a variety of fields.

There are obviously three prerequisites for the existence of isomorphisms in different fields and sciences. Apparently, the isomorphisms of laws rest in our cognition on the one hand, and in reality on the other. Trivially, it is easy to write down any complicated differential equation, yet even innocent-looking expressions may be hard to solve, or give, at the least, cumberson solutions. The number of simple mathematical expressions which will be preferably applied to describe natural phenomena is limited. For this reason, laws identical in structure will appear in intrinsically different fields. . . . However, these laws and schemes would be of little help if the world (i.e., the totality of observable events) was not such that they could be applied to. . . . The structure of reality is such as to permit the application of our conceptual constructs. . . . Yet there is a third reason for the isomorphism of laws in different realms. . . . The parallelism of general conceptions or even special laws in different fields . . . is a consequence of the fact that these are concerned with "systems," and that certain general principles apply to systems irrespective of their nature. Hence principles such as those of wholeness and sum, mechanization, hierarchic order, approach to steady states, equifinality, etc. may appear in quite different disciplines. The isomorphism found in different realms is based on the existence of general system principles, of a more or less well-developed "general system theory."[51]

Von Bertalanffy then repeats almost verbatim the distinction drawn between analogies ("worthless"), homologies, and on the third level, explanations.

General system theory can be expected to play useful roles in the meaningful transfer of models from one field to another, while weeding out the meaningless similarities. He does not specify the criteria by which general system theory will distinguish the meaningful from the meaningless similarities. Again, he indicates its special usefulness for demography and sociology and stresses that the aim of general system theory is not to replace the mechanistic view with biologism—that is, considering mental, cultural, and sociological phenomena from a merely biological standpoint. It escapes biologism by emphasizing structural isomorphies. The elaboration of a general systems theory, he maintains, will prove a major step forward in the unification of science.

The Systems Concept Applicable to the Sciences of Humanity

Psychology in many respects took its origins in a set of biases that hindered its development. Originating in the Cartesian theory of cognition, which defined subjective characters as secondary, it gradually drifted toward a behaviorist orientation. The assumption of the mind as the *tabula rasa,* having no character of its own, was the major handicap against an orientation closer to the truth. Thus, for a long time psychology was dominated by the *stimulus-response* scheme. Human and animal behavior was considered to be a response to stimuli coming from outside. Because responses are channeled through the nervous system, the brain, and the sensory apparatus, psychology was physical in its orientation.

A second major principle in psychology, derived from stimulus-response, is environmental conditioning and was developed by persons such as Pavlov, Watson, and Skinner. A third principle is equilibrium (in Freudian terms, the principle of stability); the basic function of the mental apparatus is that of maintaining homeostatic equilibrium.[52] The fourth principle at the root of modern psychology is *economy:* Behavior is governed by the principle of least effort, and environmental conditioning should be guided accordingly.

Thus, the image of man dominating psychology is that of the robot: man as a machine manufactured and trained by those using applied psychology. Its inadequacy is obvious; it disregards many aspects of behavior, including phenomena such as spontaneous play, exploratory activity, creativity, the seeking of adventure or experience beyond the horizon of the immediate environment. Behavior not only releases tensions, it also builds them up; individuals clearly react differently to the same environment. Biologically, life is not the maintenance or restoration of equilibrium, but the maintenance of disequilibrium, as the theory of the open system reveals. All original thought and all creativity are developed and expressed against the environment and at the price of expenditure of more than the minimum of energy available. Thus, the principles of thought dominating psychology, in emphasizing the dominance of the environment, the economy of action, and the principle of equilibrium speak for the worldly-wise, who by adjusting to their environment will prosper.

If life, after disturbance from the outside, had simply returned to the so-called homeostatic equilibrium, it would never have progressed beyond the amoeba which, after all, is the best adapted creature in the world—it has survived billions of years from the primeval ocean to the present day. Michelangelo, implementing the precepts of psychology, should have followed his father's request and gone into the wool trade, thus sparing himself lifelong anguish although leaving the Sistine Chapel unadorned."[53]

Anyone who wishes to originate even one single idea cannot follow the maxims of adjustment, equilibrium, and homeostasis. "Life is not comfortable settling down in pre-ordained grooves of being; at its best, it is *elan vital*, inexorably driven towards higher forms of existence."[54] True, this is a poetic simile, and metaphysical, but any image of the driving forces of the universe is such.

Against this background, a new image of man is emerging; the model of man is the *active personality system*. This is the common denominator of such otherwise different currents of thought as the psychologies of Piaget, Werner, the neo-Freudian schools, ego psychology, the theories of Allport and Maslow, new approaches in existential psychology, and others.

Thus, emphasis on the creative side of individuals, on individuality itself, on aspects that are nonutilitarian, that go beyond the immediate pressures of subsistence and survival, beyond the image of man as mere robot, are all emerging as part of the new reorientation now going on in psychology. We now recognize that man is not a passive receiver of external stimuli, but that he actively creates his universe.

THE SOCIAL SCIENCES

All social science (including sociology, economics, political science, social psychology, cultural anthropology, linguistics, history, and the humanities) is the science of social systems.[55] It will have to use the approach of general systems science. Ultimately, the largest and broadest system is that unfolded by human history. But first, system theory proves applicable to sociology; it has been adapted by Parsons, Merton, and many others. The cultural universe exemplifies systems laws.[56]

The practical applications, in systems analysis and engineering, of systems theory to problems arising in business, government, international politics, demonstrates that the approach "works" and leads to both understanding and predictions. It especially shows that the systems approach is not limited to material entities in physics, biology, and other natural sciences, but is applicable to entities which are partly immaterial and highly heterogeneous. Systems analysis, for example, of a business enterprise encompasses men, machines, buildings, inflow of raw materials, outflow of products, monetary values, good will and other imponderables; it may give definite answers and practical advice.

The difficulties are not only in the complexity of phenomena but in the definition of entities under consideration.

At least part of the difficulty is expressed by the fact that the social sciences are concerned with "socio-cultural" systems. Human groups, from the smallest of personal friendships and family to the largest of nations and civilizations, are not only an outcome of social 'forces' found, at least in primitive form, in subhuman organisms; they are part of a man-created universe called culture. . . . The cultural universe is essentially a symbolic universe.[56]

Human cultures may be approached from a system-theoretical point of view; the realm of nature is also dominated by laws which science reveals, but what about history? Does it too have laws? In von Bertalanffy's words, is a theoretical history possible? If it were possible, it would have to take a system—a culture, society, or civilization—as its unit of investigation. Although inconclusive, Von Bertalanffy is clearly hopeful that a viable systems-theory approach to history will emerge. He maintains in fact that the approaches to systematic history by figures such as Vico, Hegel, Marx, Spengler, Toynbee, Sorokin, Kroeber (the list is his), and others are serious first attempts. Whatever their differences, they all at least agree that the historical process is not completely accidental, "but follows laws and regularities which can be determined."[57]

Furthermore, the scientific approach is certainly applicable to *aspects* of society, such as in statistical approaches to population, mortality, voting behavior, and sales of products. Further logico-deductive systems approaches prove valuable in economics and in establishing a science of human values.[58]

Even for those mysterious entities, human values, scientific theories are emerging. In fact, information theory, game theory, and decision theory provide models to deal with aspects of human and social behavior where the mathematics of classical science is not applicable. Works like Rapoport's *Fights, Games, Debates* (1960) present detailed analyses of phenomena such as armament races, war and war games, competition in the economic and other fields, treated by such comparatively novel methods.

It is of particular interest that these approaches are concerned with aspects of human behavior which were believed to be outside of science: values, rational decisions, information, etc. They are not physicalistic or reductionist. They do not apply physical laws or use the traditional mathematics of the natural sciences. Rather, new developments of mathematics are emerging, intended to deal with phenomena not encountered in the world of physics.[58]

Certainly there are uncontested laws relating to immaterial aspects of culture, such as language. Linguistics has developed a number of laws covering cultural phenomena such as the mutation of consonants in the history of the Germanic languages (Grimm's laws). There are even systems phenomena in the history of art. A particular style or art form will move through a cycle from simplicity to maturity to decadence to dissolution.

Such phenomena are not disputed. The problem lies in the area of "macro-history." "Official" historians reject the notion of finding laws and regularities in this region. But if we examine some of the attempts to systematize history, such as that of Toynbee or Sorokin, and take away the metaphysics and the moralizing, what have we but conceptual models, like those of Newton in mechanics, the corpuscle or wave model in physics, the games models used to describe economic competition? The disadvantages of models are well known; but they permit one to make deductions, frame hypotheses, and predict results. This is the way theories are built. Although the "great theories" have defects, he remains confident that history itself is an unfolding of systems, whose laws will reveral themselves in this new approach.[59]

The shortcomings of the cyclic historians are natural enough in an embryonic science; instead of emphasizing these shortcomings, we would do better to emphasize their agreement in so many respects.

THE FUTURE

The future, with its looming unification of the globe by a technological civilization unified through mass media, also must be studied from a system-theoretical point of view. Spengler was right in that the West is finished and its great creative cycle is ended.

PSYCHOLOGY AND PSYCHIATRY

Psychology too is struggling away from the robot image of man; among those who have mentioned the concept of "system," von Bertalanffy cites Allport, Arieti, Buhler, Krech, Lennard and Bernstein, Menninger, Rapoport, and others. Von Bertalanffy, surveying presystem psychology, sees it as primarily authoritarian and manipulative, especially in its implications for larger societal processes. It assumed that humans, as reactive robots, were concerned only with the release of tensions and the diminution of stress.

The concept of man as robot was both an expression of and a powerful motive force in the industrialized mass society. It was the basis for behavioral engineering in commercial, economic, political, and other advertising and propaganda; the expanding economy of the "affluent society" could not subsist without such manipulation. Only by manipulating humans ever more into Skinnerian rats, robots, buying automata, homeostatically adjusted conformers and opportunists (or, bluntly speaking, into morons and zombies) can this great society follow its progress toward ever increasing gross national product.[60]

But behaviorism proved inadequate to experience. World War II for all its physical and psychological stress, did not produce an increase in neurotic or psychotic disorders; on the contrary, the affluent postwar society produced an unprecedented number of mentally ill persons. Thus, mental dysfunctions emerged that originate not from repressed drives, from unfulfilled needs, not from stress, but rather from the meaninglessness of life.

Thus, theoretical and applied psychology were led to reexamine their basic principles; now we find the conception of man as *active personality system* pervading the many different psychological schools of thought, replacing the robot model. It is to be hoped that system theory will contribute to a more adequate conceptual framework for both normal and pathological psychology. But man lives not only in a physical and physiological world; he inhabits a symbolic universe. Mental illness itself is as much a phenomenon of disturbances of symbolic functions as of physiological ones. Human behavior cannot be reduced to biologistic notions; personality disorders must now be understood in terms of the breakdown of value systems; culture is an important component of mental health.

Systems thinking provides a new conceptual framework for psychiatry. Whether we consider cognition, motivation, psychopathology, or psychotherapy, we may say that mental disease is a disturbance of the system functions of the psychophysical organism.[61]

One of the advantages of the system approach is that it is psychophysically neutral; it avoids reductionism and encompasses both the physiological processes underlying psychological events and their emotive-cognitive-cultural correlates.

THE WIDE APPLICABILITY OF SYSTEMS THINKING

Von Bertalanffy claims a wide scope for the systems concept; he specifically refers to Kuhn's studies of "revolutions" in science and regards the systems approach as the newest. As such, its scope is interdisciplinary, virtually unlimited. Consequently, we may expect to find a general systems theory and its counterpart in theoretical biology; we may also see the emergence of a set of more specialized system-theoretical disciplines, applicable to specific and narrower areas. He defines, as special cases of general system theory, the following disciplines: *applied systems research* (including systems engineering, operational research, linear programming); *computerization and simulation; cybernetics; information theory; theory of automata* (algorithmic machines that can simulate all possible logical processes); *game theory; specialized mathematical theories* (set theory, graph theory, net theory, queuing theory, decision theory); *linguistic theory* (here he makes specific reference to the linguistic theories of Whorf, who argues for the cultural relativity of linguistic concepts and categories; although his discussion is inconclusive, von Bertalanffy seems to argue that linguistic systems have counterparts in the physiological and sensory organizations of the beings who carry them).[62]

GENERAL UNITY OF VON BERTALANFFY'S THOUGHT

Von Bertalanffy's general world view and his claims for universal significance of system theory emerge early in his writings, which are rather repetitious and even static in character; many arguments are repeated almost verbatim from one book to another, which at least makes his views consistent and easy to summarize. Thus, his other general writings, such as his *Organismic Psychology and Systems Theory*[63] and the paperback version thereof, entitled *Robots, Men and Minds—Psychology in the Modern World*[64] are virtually identical, apart from a few references to topical writings and issues.

Von Bertalanffy thus offered a programmatic philosophy based on biology; he saw himself as standing at the beginning of an important new development, but looked to others to develop and extend his concepts into specific specialties. Such was the program of the Society for General System Research. The elaboration of his views into a more systematic philosophy has been undertaken by a number of his associates. Before exploring the elaboration of systems theory into a philosophy and its migration into other disciplines, we must first explore developments in other areas which originated independently of his work.

COMMENT

Systems theory can be seen in its social philosophy as another variant of *organicism*, the image of society as an organism subject to "growth" and "decay," which "evolves" over time into new and more differentiated forms. This image, of course, can be, and has been, developed further: Societies can be seen as rival species competing for survival and as "evolving" in the process. In addition, various social institutions will be likened to organs of the body; perhaps the military will be seen as the teeth and claws of the social organism, the university as brain, the mass media as senses, and the like.

In this work system theory is viewed as one more variant of this organicism, its resurrection in the twentieth-century dress of cybernetic terminology. Systems theory presents some aspects that are new—emphasis on computers, cybernetics, information theory, operations reserach, and the like, many of whose concepts are specific to twentieth-century science and technology— and other aspects that are quite old—in applying systems concepts to social matters, systems thinkers fall into the organicist conception of society; many do so with little or no awareness of the great age of this imagery.

Organicism as such is an old way of conceiving of society. We might also remark, following Kenneth Bock, that evolutionist views of society precede the work of Darwin and what is called "social Darwinism." For further discussion, see the following: *The Acceptance of Histories* (Berkeley: University of California Press, 1956); Robert A. Nisbet, *Social Change and History* (New York, Oxford University Press, 1968); Kenneth Bock; "Evolution, Function, and Change," *American Sociological Review*, (April 1963), pp. 229–237); Kenneth Bock, "Darwin and Social Theory," *Philosophy of Science*, Vol. 22 (1955), pp. 123–134). I am indebted to Stanford Lyman for helpful suggestions on this point.

NOTES

1. Thomas S. Kuhn, *The Structure of Scientific Revolutions* (Chicago: University of Chicago Press, 1920).
2. Ludwig von Bertalanffy, *General System Theory: Foundations Development Applications* (New York: Braziller, 1968, pp. 187–188.
3. Stephen C. Pepper, *World Hypotheses—A Study in Evidence* (Berkeley and Los Angeles: University of California Press, 1942, reprint ed., 1970.
4. *Ibid.*, pp. 21ff.
5. *Ibid.*, pp. 243–245.
6. *Ibid.*, p. 271.
7. See E. E. Emery, ed., *Systems Thinking* (Baltimore: Penguin, 1969), pp. 57ff.
8. Pepper, *op. cit.*, pp. 280ff.

9. *Ibid.,* p. 307.
10. *Ibid.,* p. 310.
11. *Ibid.*
12. Most of Henderson's works on sociology and on the philosophy of science have been collect-ed, edited, and introduced by Bernard Barber, who offers a thorough discussion of Henderson's work and main ideas. See *L. J. Henderson on the Social System: Selected Writings,* ed. and with an Introduction by Bernard Barber (Chicago, University of Chicago Press, 1970).
13. "Such assertions then are non-logical, and neither logical nor illogical. According to Bridgman's definition of meaningless, *reality* is meaningless because no operation can be agreed upon as a definition of the word reality." *Ibid.,* p. 167).
14. *Ibid.,* p. 118, also p. 189.
15. *Ibid.,* p. 27.
16. *Ibid.,* p. 28.
17. *Ibid.,* pp. 136–139.
18. *Ibid.,* p. 88.
19. *Ibid.,* pp. 188–189.
20. Walter Bradford Cannon, *The Wisdom of the Body,* originally published 1932; revised and enlarged edition, 1939; see reprint (New York: Norton, 1963).
21. *Ibid.,* pp. 305ff.
22. *Ibid.,* p. 308; the material here summarizes his chapter, in detail.
23. *Ibid.,* p. 314.
24. *Ibid.,* p. 323.
25. In addition to this wide-ranging analogy of organism with society, Cannon offered one other narrower concept that proved important for systems theorists—the notion of *internal environ-ment:* Simply, the internal cells and organisms of the body do not for the most part cope with the external environment; nevertheless, they interact with the blood stream as an environ-ment.
26. In the following discussion we present a summary of the main ideas of von Bertalanffy, rather than a detailed survey of his writings. Since his philosophical ideas, as well as his views of society, emerge at an early date in his writings, a chronological survey of his writings is less appropriate than a discussion in terms of his leading ideas.
27. Ludwig von Bertalanffy, *Modern Theories of Development: An Introduction to Theoretical Biology,* trans. and adapted by J. H. Woodger (London: Oxford University Press, 1933). *Problems of Life—An Evaluation of Modern Biological Thought* (New York: Wiley; London: Watts, 1952).
28. F. E. Emery, *Systems Thinking* (Baltimore: Penguin Books, 1969), p. 57.
29. These views can be found in essentially the same form in most of his publications before and since; see especially his book, *General System Theory: Foundations, Development, Applications* (New York: Braziller, 1968).
30. *Ibid.,* p. 60. He here and elsewhere argues that scientific laws exist on multiple levels. The first level is the laws of physics; the second level, the laws of biology. The quotation above continues: "A third level of statistics would probably be required for sociology. The exact treatment of this problem will presumably require totally new forms of logico-mathematical treatment."
31. *Ibid.,* p. 65.
32. *Ibid.,* p. 119.
33. *Ibid.,* p. 186.
34. Von Bertalanffy, *Problems of Life,* p. 172.
35. *Ibid.,* p. 201.
36. *Ibid.,* pp. 200–201.
37. In *Science,* Vol. 111 (1950), pp. 23–29; reprinted in Emery, *op. cit.,* pp. 70–85.

38. *Ibid., passim,* The following is a close paraphrase of von Bertalanffy's essay.
39. In Emery, *op. cit.,* p. 82.
40. "General Systems Theory—A Critical Review," Vol. 7 (1962), pp. 1–20. Reprinted in Walter Buckley, ed., *Modern Systems Research for the Behavioral Scientist* (Chicago: Aldine, 1968), pp. 11–30.
41. *General System Theory,* pp. 20–21.
42. Von Bertalanffy has seemed somewhat anxious to establish his priority in first formulating the concept of the open system, and also the concept of system theory as a new science.
43. *General System Theory,* p. 38.
44. *Ibid.,* pp. 54ff.
45. *Ibid.,* pp. 60ff.
46. *Ibid.,* p. 62.
47. *Ibid.,* p. 62.
48. *Ibid.,* pp. 62–63.
49. *Ibid.,* pp. 80ff.
50. *Ibid.,* pp. 80–81.
51. *Ibid.,* pp. 82–84.
52. *Ibid.,* pp. 188ff. The material presented here is a close summary of his views which are expressed in virtually identical terms throughout all his writings.
53. *Ibid.,* p. 192.
54. *Ibid.,* p. 192.
55. *Ibid.,* pp. 194–195.
56. *Ibid.,* pp. 196–197.
57. *Ibid.,* p. 198.
58. *Ibid.,* pp. 198–99.
59. *Ibid.,* pp. 200–201.
60. *Ibid.,* p. 206.
61. *Ibid.,* p. 26–8.
62. For his claim that general system theory has counterparts in specific fields see *ibid.,* chapter 1; for his discussion of Whorf, see *ibid.,* chapter 10.
63. Clark University Press with Barre Publishers, Barre, Mass., 1968.
64. New York, Braziller, 1968.

Cybernetics

M ost practioners of cybernetics offer claims to sociological, philosophical, and religious significance beyond the limits of a specific technique or discipline. Norbert Wiener, the founder of cybernetics, argued that this science would have an immense impact on civilization, not merely in the form of automating a great variety of functions previously performed by human labor, but even more through its philosophical import. Wiener argued that cybernetics in its full implications had much to offer even for theology, affecting our conceptions of God and humanity.

The major theoretician of cybernetics after Wiener is W. Ross Ashby; both men, in addition to purely technical work in elaborating the findings and theorems of cybernetics, have written missionary works explaining the nature of the field and its wider importance. Their writings are complementary: Ashby writes more as a popularizer of basic cybernetic concepts, and Wiener wrote either highly technical material or books interpreting its significance for a lay audience.

We are not concerned here primarily with the technicalities of cybernetics, but rather with its social and philosophical claims. Nevertheless, some of its intrinsic content is important because its imagery and terminology have been adopted by writers on social matters. Therefore, we examine here only enough of the basic concepts of cybernetics, as propounded by Ashby and Wiener, as is relevant to their social claims.

W. ROSS ASHBY

Ashby's major missionary works are *An Introduction to Cybernetics* and *Design for a Brain*. Their major purpose is to account for change and behavior that appears to show "purpose," "memory," "foresight" in purely determinist and mechanistic terms. Ashby constructs a set of concepts derived from cybernetics in order to do so.[1]

The basic concept of cybernetics is patterns of change; the basic patterns must be classified and appropriate concepts must be derived.

OPERANDS AND TRANSFORMS

If a set of states $(a, b, \ldots)$ is acted upon, it may move to another set called the transform. A set of transitions is called a *transformation*. If a set of transforms contains no states or terms not found in the set of operands (i.e., contains no new states), the set is said to be *closed* under the transformation. A transformation may be *single-valued* (all states converge to one), or it may be *one-one* (all the transforms differ from one another.).

A *determinate machine* is one that behaves the same as a closed single-valued transformation, and cybernetics deals with ways of behaving that are determinate, that follow regular and reproducible courses. Cybernetics studies the determinateness of ways of behaving, not the material substances that do the behaving.

Such a system may go through a sequence of states that is called a *trajectory* or a *line of behavior*. A determinate machine is one that from any one state cannot proceed to both of two different states; it must always proceed to one and the same state from the initial state. Thus, the cybernetic approach is based on the concept of the correspondences between transformations and real systems. When a machine can be fully represented by a transformation, the transformation is said to be a *canonical representation* of the machine, and the machine is said to embody the transformation.

Cybernetics is based on the assumption that complex systems can be studied by monitoring the state of the machine taken as a whole, without detailed specifications of the parts, and that a theory of unanalyzed states can be rigorously developed. If more detailed analysis is required, knowledge can be expressed in a list of the states taken at any moment by each of its parts. Such a list of quantities is called a *vector*, a compound entity having a definite number of components; this is a variable, but one that is more complex than the kind met with in elementary mathematics.

Once the transformations of the machine are known, the set of equations so obtained is the canonical representation of the system; from this, all descriptions of a determinate dynamic system may be developed.

Thus, given a transformation and an initial state, one can always develop the trajectory of a determinate machine; if the equations developed are described as nonintegrable or solvable, the trajectory cannot be obtained, if one is restricted to certain defined mathematical operations. But investigators have shown how equations of the unsolvable type can be studied and in practice usable results can be obtained. Further, the study of the transformations of a machine can be helped by geometric demonstrations where purely albegraic forms do not clarify the results.[2]

Ashby indicates that cyberneticists realize the material object under study is not the system, because every material object contains an unlimited number of variables and, therefore, of possible systems. The system is a list of variables.[3] The most common task of the experimenter is that of varying his list of variables until he finds the set that gives the required singleness of value and, therefore, determinateness.

Thus *the determinate machine is one whose behavior can be encompassed in a list of variables that is logically and mathematically workable.*

THE MACHINE WITH INPUT

Since every machine may be acted upon by various conditions, the task is to define what in the symbolic world of transformations corresponds to this. Consider a machine with a lever that could be put in any of three positions; a different way of behaving may correspond to each position. Thus there are two forms of change:

1. Change from one state to another within a way of behaving.
2. Change from one way of behaving to another.

Instead of considering each transformation separately, a set of transformations can be regarded as a transformation itself (considering, however, only terms that are closed and single-valued). A symbolic designation of a way of behaving is called a *parameter.*

A real machine whose behavior can be represented by a set of closed single-valued terms is referred to as a *transducer* or a *machine with input.* The set of transforms is its canonical representation; the parameter, as something that can vary, is its input. These too can be represented mathematically.

INPUTS AND OUTPUTS

In machines and other determinate physical systems, inputs and outputs may be obvious or restricted to a few physical terminals. But in biological systems the number of parameters is usually very large, and the whole set is not always obvious. *Inputs* may refer to the few parameters appropriate to a simple mechanism or the many parameters appropriate to the free-living organism in a complex environment. But according to Ashby, the theory of such inputs can be as rigorous. In the relation between an organism and its environment, inputs to one may be outputs to another, depending upon the point of view: Thus, when an organism interacts with its environment, its muscles may be conceived of as the environment's input and its sensory organs as the environment's output.

COUPLED SYSTEMS AND COUPLING WITH FEEDBACK

Determinate machines can be coupled to one another; the inputs and outputs of the machines are linked to one another. If both machines are determinate in the sense defined above, the two machines will form a new machine of completely determinate behavior. The two machines can be coupled so that each affects the other. The new relations derived can also be expressed in mathematical terms. When machine A affects machine B but not vice versa, we say "A dominates B." But cybernetics is especially interested in the case where each affects the other. Where this exists, "feedback" is said to be present. The concept of feedback becomes inadequate when applied to complex dynamic systems where the interactions are too complex to determine unequivocally what is affecting what. But what is important is that complex systems that have rich internal cross-connections show complex behaviors and that these behaviors can be goal-seeking in complex patterns.[4]

Effects between machines can be over one step (immediate effects) or over several steps—for example when one machine affects a third variable which in turn affects the second machine (ultimate effects). Simple nonmathematical diagrams can determine which variables dominate others.

Systems may also show the important property of *reducibility:* under some conditions the effects of one subsystem on another may be reduced or eliminated to the point where the subsystems are virtually independent.

Using these basic concepts as a starting point, cybernetics considers the very large system such as (in Ashby's terms) a computer, a nervous system, or a society. In cybernetics size does not mean mass. It means *complexity,* the number of distinctions that have to be made, the number of states available, or the number of components in a vector. "Very large" means too complex to observe or control completely. But size does not invalidate the theorems and principles developed so far; it may make their applicability to a particular case problematic.

RANDOM COUPLING

Given a complex system, the observer can specify it only incompletely—that is, statistically. Since any subsystem could be coupled to any other, random methods of choosing couplings are used. What results is not a machine but a set of machines. Thus although a system may be too large for its details to be specified individually, it may be specified statistically.

If we consider a system of a given degree of largeness, all of whose parts are identical and with enough coupling between the parts to give the system coherency, there may exist a small number of immediate effects. But the effects may also be due to the fact that one part or variable affects another only under certain conditions so that the immediate effect is present for much of the time in only a nominal sense. Thus a variable may spend much of the time not varying. One common cause is the existence of a threshold below which a variable shows no change except when the incoming disturbance exceeds a definite value—for example, "the voltage below which an arc will not jump a given gap, and the damage that a citizen will sustain before he thinks it worth while going to law. In the nervous system the phenomenon of threshold is, of course, ubiquitous."[5]

The existence of threshold may be understood as a phenomenon that cuts a whole into temporarily isolated subsystems. The higher the level of threshold, the more isolated the parts become.

LOCAL PROPERTIES

Large systems with much repetition of parts, few immediate effects, and slight couplings may show much *localization* of effects so that the occurrence of a property in localized form does not determine its occurrence elsewhere in the system. This is often the case in chemical systems where ions may affect only a few immediate neighbors and have little or no effect on the majority of other ions; but in biological systems the interconnections may be so great as to minimize localization.

PROPERTIES THAT BREED OR ARE SELF-LOCKING

Systems may in time develop properties that resist reversal: Chemicals in a solution may form an insoluble subcompound that goes out of solution and is irreversible; in the cerebral cortex some circuits may develop reverberations that are too strong for suppression by inhibition; or workers in an unpleasant industry may, after a stretch of unemployment, go into more pleasant forms of employment and then refuse to return. Such self-locking changes are generally important in determining the eventual state of the system.

Parts that show a given property may induce other parts affected by them to show that property. A property may "breed" slowly (an epidemic) or rapidly (an explosion, the spread of a fashion, or an avalanche). If the breeding rate of a property can be expressed in a time equation, depending on the key values, the property may decline with time, remain constant, or increase. Whether such properties exist in the cerebral cortex is unknown; but if they do, they will be of outstanding importance and will impose outstanding characteristics on the cortex's behavior. Ashby adds, "It is important to notice that this prediction can be made without any reference to the particular details of what happens in the mammalian brain, for it is true of all systems of the type described." [6]

The theory of very large systems is not very different from the elementary systems described earlier; its construction is merely a matter of time and labor rather than of any profound or peculiar difficulty.

Another major concept in understanding cybernetics is that of *stability*. The concept carries several meanings.

1. *Invariant.* Some aspect of a system that is unchanging although the system as a whole passes through a series of changes.

2. *Equilibrium.* At certain values a transformation does not cause the operands to change (e.g., there are differential equations that are stable with respect to time).

3. *Cycle.* A transformation may take a representative point of a system repeatedly around a sequence of values.

4. *Stable region.* If variable b goes to variable g while g goes to b for each step of the transformation, no new state is generated.

A *disturbance* is defined as something that moves a system from one state to another. Small disturbances are in practice constantly acting on systems. A system may be stable if the disturbance lies within a certain range.

Stability, Ashby remarks, is not always to be considered desirable. He cites the examples that if high-IQ persons fail to reproduce, average IQ might stabilize at a low average. Similarly, in neurology damaged nerves are often in a two-step condition—either "quiet" or extremely painful.

He also speaks of subsystems having a power of veto: A system consisting of coupled subsystems is in a state of equilibrium only if each part is at a state of equilibrium in the conditions provided by the other parts; no state of the whole can be a state of equilibrium unless it is acceptable to every one of the component parts. The ideas involved in the concept of stability can be summarized.

1. *Equilibrium* is a state unchanged by a transformation.

2. *Multiple equilibrium* is a repeating cycle within which a system may stabilize.

3. *Stability* exists against all disturbances within a certain range of values.

4. *Stability/instability* in a large system may be described by terms such as: "will return to its usual state after a disturbance," or "will show ever-increasing divergence after a disturbance."

THE BLACK BOX

The concepts described above are preliminary to the next step in cybernetics in which the "black box" is considered. The term is used in electrical engineering. An engineer is given a sealed box with terminals for input and output; he may apply any voltages he pleases and may observe what he can; he is to deduce what he can of the contents of the black box.

The problem is applicable in daily life. We constantly manipulate such objects; a brain or an animal may be regarded as a black box. Only by accepting the casing on the black box can we develop a scientific epistemology.[7] We begin by making no assumptions about its contents, but we have certain resources for acting on it. Let us assume that the inputs can be represented by a set of levers or pointers and that the output is represented by a set of dials. Though it may seem unnatural, this representation can represent the great majority of natural systems, even biological or economic systems, according to Ashby.

THE INVESTIGATION

The system is investigated by a protocol, a long list, drawn out in time, that shows the sequence of input and output states. It is a sequence of values of the vector that has two components; input state and output state. All knowledge obtainable from a black box is what can be obtained by recoding the protocol. No skill is needed in manipulating the input; since nothing is known of the box, no facts exist that justify any particular method. When a generous length of record is obtained, the experimenter can demonstrate that the behavior of the box is machine-like, and he can deduce its canonical representation. If the machine is not determinate (i.e., if a transformation is not single-valued), he can proceed in either of two ways:

1. He may alter the set of inputs and outputs to take more variables into account.

2. He may look for *statistical* determinacy (regularity of averages, etc.) instead of strict determinacy.

States may appear that are inaccessible—that is, once past, they cannot be made to appear again (e.g., a mine once exploded or an animal once it learns a task). Sometimes, if more states are added to the transformations, a previously inaccessible state may be made to reappear.

DEDUCING CONNECTIONS INSIDE THE BLACK BOX.

When an adequate protocol has been accumulated, something can be deduced about the internal connections in the black box. But it has been shown that any given behavior exhibited by a black box can be produced by an infinitely large number of possible networks of inner connections. The behavior does not specify the connections uniquely.

ISOMORPHISMS

A map and the countryside it represents may be *isomorphic*. So may a moving object and an equation and a photographic negative and a print obtained from it. Other isomorphisms include a machine of mechanical nature and an electrical device and a certain differential equation—all three systems may be isomorphic. Thus an electrical device may be a "model" of a differential equation, an analogue computer. "The big general-purpose digital computer is remarkable precisely because it can be programmed to become isomorphic with any dynamic system whatever."[8]

Isomorphic devices are valuable in science. One form may be workable over a patch in which the other is difficult to manipulate. The concept of isomorphism can be shown to be capable of exact and objective definition. The canonical representations of two machines are isomorphic if a one-one transformation of states of one machine to the other can convert the one representation to the other. But the relabeling may take various levels of complexity; transformations may not be simple, but complex.

HOMOMORPHISMS

Homomorphisms occur when a many-one transformation, applied to a more complex machine, reduces it to an isomorphism of a simpler machine. Two machines are "equal" only when they are so alike that an accidental interchange would be undetectable by any test applied to their behaviors. There are also lesser degrees of resemblance—for example, two identical systems operating on different time cycles. If a machine can be subject to a many-one transformation so that the transform is isomorphic to another machine, the two are "homomorphic." Biological systems are too complex for such simple transformations; but if a complex system's states are compounded suitably, it may be simplified to a new form.

One must deliberately refuse to attempt all possible distinctions; study must be restricted to a homomorphism of the whole. Therefore, knowledge of a system need not be exhaustive; partial knowledge over the whole may be sufficient for practical purposes.

The term *system* may refer to the thing in itself, but also may refer to the set of variables with which an observer is concerned. Science, according to Ashby, is not concerned with discovering what the system "really is," but with coordinating observers' discoveries, each of which is only a portion of the truth.

LATTICES

Systems may be mathematicized further by recourse to lattice diagrams. A machine may be simplified in a series of logical steps by *merging* the subsystems within the machine. By examining all possible combinations of mergings, one obtains a graded set of simplifications that range from the full machine in all its complexities to the machine with subsystems *a* and *b* merged, *a* and *c* merged, *a* and *b* and *c* merged, and so on. The diagram of such a set of all forms of a machine and its subsystems and their interconnections is called a lattice, a much-studied structure in modern mathematics. Much that has previously been learned only intuitively can be learned about systems by the study of lattices. Ashby considers this method applicable to machines and economic and social systems. Lattices, he suggests, even generate concepts. Thus the state in which all distinctions have been merged corresponds to the idea of persistence. One can say only that the machine persists and nothing more. Ashby does not clarify why the word *persistence* predates cybernetics.

MODELS

A model is seldom isomorphic with a biological system; usually it is a *homomorphism,* or two systems—say, a biological system and a model—are so related that a homomorphism of one is isomorphic with a homomorphism of the other. This is a "symmetric" relation; each is a "model" of the other.

Properties ascribed to machines can also be ascribed to black boxes. Ashby tells us that we work with black boxes often in our everyday lives. For example, we ride a bicycle with no knowledge of the interatomic forces that hold the metal together. Real objects are black boxes, and we have been operating with them all our lives. "The theory of the Black Box is simply the study of the relations between the experimenter and his environment, when special attention is given to the flow of information."[9] Adopting a phrase from information theory, Ashby suggests that the study of the real world becomes a study of transducers.

EMERGENT PROPERTIES

In studying a number of black boxes in isolation until each one's canonical representation is established, if they are coupled in a known pattern by known linkages, it will follow that the behavior of the whole is determinate and can be predicted. "Thus, an assembly of Black Boxes, in these conditions, will show no 'emergent' properties; i.e., no properties that could not have been predicted from knowledge of their parts and their couplings."[10]

Some examples of emergent properties given by Ashby include the following:

1. Ammonia is a gas, as is hydrogen chloride; when they are mixed, the result is a solid.
2. Carbon, hydrogen, and oxygen are all practically tasteless, but their compound, sugar, has a sweet taste not possessed by any of them.
3. The 20 and more amino acids in a bacterium are not self-reproducing, but the bacterium is.

Properties emerge, Ashby tells us, when knowledge is incomplete, and therefore biological systems often show "emergent" properties. The larger and more complex the system, the more laborious is the study of its features. Thus Newtonian physics has in principle solved all gravitational problems; yet its application to three bodies is very complex, and to half a dozen is prohibitively laborious. Yet astrophysicists want to ask questions about the behavior of star clusters with 20,000 members. How can this be done? The physicists, led by Poincaré, have now a well-developed method for dealing with such matters—topology. Similar approaches, applied to complicated differential equations, enables the main important features of the solution to be deduced in cases where full solutions would be unmanageable—the so-called "stability" theory of these equations. In the 1930s Kurt Lewin attempted a topological psychology, but then topology was not yet fully developed. In the 1950s, it was much better developed, especially by the French school of mathematicians publishing under the pen name of N. Bourbaki. Ashby indicates that we now have the possibility of a rigorous and practical psychology.

THE INCOMPLETELY OBSERVABLE BOX

An important area of cybernetic theory involves a box, some of whose "dials" are missing—for example, a biological organism or the brain. Its behavior may seem miraculous when all significant variables are not observable. "It is possible that some of the brain's 'miraculous' properties—of showing 'foresight,' 'intelligence,' etc., are miraculous only because we have not so far been able to observe the event in *all* the significant variables."[11]

Ashby illustrates the point by a system completely observable to one observer and only partially observable to another. In the example the partial observer can completely record system relations after a time lag; he restores predictability by taking the system's past history into account; he assumes the existence of some form of "memory." But the possession of "memory" is not a wholly objective property of a system; it is a relation between a system and an observer, which will alter if the channels of communication between the system and the observer change. His illustration, fairly representative of the type of illustration, is as follows:

The observer is assumed to be studying a Black Box which consists of two interacting parts, A and Z. Both are affected by the common input I. (Notice that A's inputs are I and Z.)

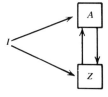

Suppose the important question is whether the part A does or does not show some characteristic behavior B [i.e., follow trajectory B]. Suppose this is shown (followed)

only on the simultaneous occurrence of (1) I at state α and (2) Z at state y. Suppose that Z is at state y only *after* I has had the special value.

We (author and reader) are omniscient, for we know everything about the system. Let us, using full knowledge, see how two observers (One and Two) could come to different opinions if they had different powers of observation.

Observer One can see, like us, the values of both A and Z. He studies the various combinations that may lead to the appearance of B, and he reports that B appears whenever the whole shows a state with Z at y and the input at α. Thus, given that the input is at α, he relates the occurrence of B to whether Z is at y now.

Observer Two is handicapped—he can see only I and A, not Z. He will find that knowledge of A's state is not sufficient to enable him to predict reliably whether B will be shown (for sometimes Z will be at y and sometimes at some other state). If however Two turns his attention to *earlier* events at I he finds he can predict B's appearance accurately. For if I has in succession the values μ, α then behavior B will appear, and not otherwise. Thus, given that the input is at α, he relates the occurrence of B to whether I did have the value earlier.

Thus Two, being unable to observe Z directly, can none the less make the whole predictable *by taking into account earlier values of what he can observe*. The reason is, the existence of the correspondence:

I at μ earlier ——— Z at y now
I not at μ earlier ——— Z not at y now

As this correspondence is one-one, information about I's state a step earlier and information about Z's state now are equivalent, and each can substitute for the other; for to know one is to know the other.

If One and Two are quarrelsome, they can now fall into a dispute. One can maintain that the system shows no "memory," i.e., its behavior requires no reference to the past, because the appearance of behavior B can be fully accounted for by the system's *present* state (at I, A and Z). Two can deny this, and can point out that the system I and A can be shown as determinate only when past values of I are taken into account, i.e., when some form of "memory" is appealed to.

Clearly we need not take sides. One and Two are talking of different systems (of I + A + Z or of I + A), so it is not surprising that they can make differing statements. What we must notice here is that Two is using the appeal to "memory" as a substitute for his inability to observe Z.

Thus we obtain the general rule: *If a determinate system is only partly observable, and thereby becomes (for that observer) not predictable, the observer may be able to restore predictability by taking the system's past history into account, i.e., by assuming the existence within it of some form of "memory." . . . Thus the possession of memory is not a wholly objective property of a system—it is a relation between a system and an observer; and the property will alter with variations in the channel of communication between them.*

Thus to evoke "memory" in a system as an explanation of its behavior is equivalent to declaring that one cannot observe the system completely. The properties of "memory" are not those of the simple "thing" but the more subtle "coding." [12]

Ashby concludes that memory is not an objective something that a system does or does not possess, but is rather a concept invoked by an observer to fill in a gap when the system is not completely observable. Memory in the brain is only partly objective. "No wonder its properties have sometimes been found to be unusual or even paradoxical. Clearly the subject requires thorough reexamination from first principles." [13]

The weakness of this argument should be obvious. It begs the question by implicitly defining memory as an epiphenomenon of mechanized systems.

The exposition of Ashby's views presented above represents an outline of basic concepts. Once these are established, Ashby develops them in terms of their wider applications. Determinate machines, whether systems or organisms, exhibit various types of regulation and control. To examine these we must explore the concepts associated with variety of behavior.

VARIETY

In discussing regulation and control in mechanisms, one must examine *sets of possibilities* to determine what a machine might do under these various sets. Science must proceed in this way because its focus is not on the unique but on the repetitive. Popular language is confusing in this respect; we say "the chloride ion" when we mean "the set of all chloride ions." Probability statements refer to the set of events and not to the individual. The concept of the set plays a major part in the communication theory of Shannon and Wiener. *Variety* refers to either (a) the number of distinct elements in a set or (b) the logarithm to base 2 of this number. In the second form this is called a binary digit, commonly abbreviated to *bit.*

Constraint

Under certain conditions variety may be less than under others; not all combinations of possibilities in a set are used. Constraint is an especially important concept in that when a constraint exists, advantage can usually be taken of it. Shannon's work on communication theory is especially based on constraints. Defined by Ashby,

Constraint is a *relation* between two sets, and occurs when the variety that exists under one condition is less than the variety that exists under another. Thus, the variety in the human sexes is 1 bit; if a certain school takes only boys, the variety in the sexes within the school is zero; so as 0 is less than 1, constraint exists. . . . Similarly, constraints may be slight [i.e., they eliminate few possibilities] or they may be severe [eliminating many].[14]

Objects as Constraints

A chair is a collection of parts, but their relative positions are subject to constraints. The chair's existence as an object, a unity, corresponds to the presence of constraint. A world without constraints would be a flowing chaos. Constraints make prediction possible; thus an aircraft in flight is subject to constraints in its trajectory, which makes predicting its path possible. The concept of a machine, as developed from inspecting its protocol, comes from recognizing that the sequence in the protocol shows some form of constraint. One important constraint is that of continuity wherein a value can change only to a neighboring value.

Learning and Constraint

Learning can occur only under conditions of constraint, otherwise the environment is too *fluid*. For example, a rat exploring a maze could never learn the way through if the structure of the maze were continually changing.

Variety in Machines

Under some transformations variety may decrease; machines may "settle" into ways of behaving that are less than the total set of possibilities. Given a set of states and a single-valued transformation, we can predict that as time progresses variety in a set cannot increase and will usually diminish. Thus the observer's uncertainty about the state of a system can only diminish. The *law of experience* indicates that, given a set of machines with a maximum variety, changes of parameter values cannot increase the set's variety; the longer a routine, the more the set will approach a standard state.

COMMUNICATIONS THEORY

At this point cybernetics approaches communications theory: The variety of input and output between machines can be regarded as a form of communication. Ashby illustrates by reference to a broadcast of a speech. Vibrations undergo vast metamorphoses from speech to reception—variety of printed units on a page; variety of sounds formed by the speaker's lips, tongue, and so on; variety of compression waves in the air; variety of vibration patterns in the microphone; variety of electrical impulses in the amplifier, and so on. Through all of this, something is preserved. Decoding is the process of applying a transformation to supply the original message. If variety of messages are to pass through several codes in succession and are to be uniquely restorable to their original forms, the process must be one that preserves the variety in the set at every stage.

Communication theory is largely concerned with the processes by which messages (variety) can be coded, transmitted, and decoded. Variety can be transmitted through a small intermediate transducer, given sufficient time. Thus shrinkage in channel capacity can be compensated by an increase in sequence length. For example, the variety of the English alphabet is greater than the telegraphic distinction of dot-and-dash; but the latter, by combinations of enough patterns of dots and dashes, can transmit the larger variety of letters.

Interference of Messages

If acid and alkali are passed down the same pipe, they destroy each other. What happens to two messages in the same channel? They can be carried without mutual destruction. For example, two messages can be carried by a set of letters; the first is conveyed by the letters themselves; the second, in binary form, can be conveyed by distinguishing capitals from lower case letters. The variety in the received form of the message must be no less than the variety of the original. Provided that no variety is lost and that the mechanism is determinate in all its details, the two messages continue to exist. All that is necessary is a suitable converter, and its construction is always possible. All that is required is that the variety in the received forms must be not less than that of the original. The fact that chaos does not necessarily occur when two or more messages are carried in the same channel is of special importance in neurophysiology, especially in the study of the cerebral cortex. How the cortex prevents the destructive interaction and chaos that might result from the multiplicity of incoming messages is not at present clear.

Cybernetics and Communication Theory

Ashby, like most cyberneticists, regards cybernetics and the mathematical theory of communication developed by Shannon and Weaver to be closely related, and in some cases, virtually identical.

There are some channels, such as nerves and telephone cables, that transmit information that is sustained for an indefinitely long time. This type of transmission has been the special province of communications theorists such as Shannon and provides a transition point from the closed discontinuous changes described so far to the more complex probabilistic processes that are the ultimate concern of cybernetic theory.

The single-valued determinate transformations described in the elementary exposition of cybernetic theory are a special and extreme case of the stochastic transformation. Ashby illustrates as follows. Consider the following matrix of transitions:

	A	B	C
A	0	1	0
B	1	0	0
C	0	0	1

This represents a single-valued transformation:

A	B	C
B	A	C

Thus A always moves to B, B to A, and C always remains C. Now, if these values are replaced by probabilities, we might have something as follows:

	A	B	C
A	0	0.9	0.1
B	0.9	0	0.1
C	0	0.1	0.9

Now A moves to B 90% of the time, and to C 10% of the time, and so on. We now have a matrix of transition probabilities. Ashby here adopts the "classical" definition of probability, based on long-term trials and the outcomes of large numbers of trials; thus he adopts Ronald Fisher's definition of a probable event as a frequent event; this is based on the assumption that events exhibit a stable and constant probability over long period of time.[15]

But the question now arises: What would a system behave like if its transformations were not closed and single-valued, but stochastic? Ashby uses the illustration of an insect living near a shallow pool that spends some time in the water, some time under pebbles, and some time on the bank. He suggests that a complete protocol of past behaviors of the insect could be kept so that one could record the proportions of the insect's time that is spent in each of the three areas and, further, of the number of times when at any one place— for example, in the water—it moves to the pebbles against the number of times it moves from the water to the bank.

Thus a matrix of transition probabilities is developed. Where these are stable over time, it is known as a Markov chain.

If we now consider a large number of insects living at this pond, each assumed to behave independently of the others (a doubtful assumption), we lose sight of the individual insects and see three populations, one on the bank, one in the water, one under the pebbles.

Ashby stresses that the system composed of the three populations is determinate, although the individual insects behave only with certain probabilities.[16]

If we experiment with the insect population by forcing 100 of them under pebbles and then watching what happens, we see that in the long run the system settles down, in a series of dying oscillations, to a state of equilibrium.

It is worth noticing that when the system has settled down, and is practically at its terminal populations, there will be a sharp contrast between the populations, which are unchanging, and the insects, which are moving incessantly. The same pond can thus provide two very different meanings to the one word "system." ("Equilibrium" here corresponds to what the physicist calls a "steady state.")[17]

The Markov chain as defined above carries with it one other important feature: "The probabilities of transition must not depend on states earlier than the operand."[18]

"Thus if the insect behaves as a Markov chain it will be found that when on the bank it will go to the water 75% of the cases, whether before being on the bank it was at bank, water, or pebbles."[19]

Many sequences are not Markov chains; the system's "memory" seems to extend back for more than one state. Language, says Ashby, is especially replete with instances; thus in English the probability that a given letter, say s, will be followed by t depends upon what preceded the s and will vary accordingly.[20]

Non-Markovian systems can often be redefined to produce a Markov chain, if the transition probabilities depend in a constant way on the states preceding the operand. This is done by defining new states as vectors over n steps, assuming that a suitable n can be found.[21]

Entropy

A fundamental concept in communication theory and in cybernetics is *entropy,* as defined by Shannon. It is primarily a measure of variability and as such is related to the statistical concept of sampling variability.

Any set of probabilities (i.e., a set of positive fractions whose sum is 1) is associated with the following sum:

$p_1 \log p_1 - p_2 \log p_2 - \ldots - p_n \log p_n$

of the set of probabilities and is designated as H. (The logarithms can be calculated either to base 10 or base 2; in the latter case, the sum is the number of bits.) Since the probabilities are less than 1 (except in the special and extreme case where one event has probability of 1 and all others have probability of zero, in which case the variety—and therefore the entropy—is zero), each log thereof is a negative quantity; multiplying bby minus the corresponding probability therefore produces a positive sum. Ashby uses the following simple illustration: A set of traffic lights having a variety equal to 4 is given: (1) red (stop); (2) red and yellow (prepare to go); (3) green (go); (4) yellow (prepare to stop). Assume that these are on for durations of 25, 5, 25, and 5 seconds respectively, so that a motorist who turns up at random times would find the lights in various states with corresponding probabilities of 0.42, 0.08, 0.42, and 0.08. Their entropy then is

$-0.42 \log 0.42 - 0.8 \log 0.08 - 0.42 \log 0.042 - 0.08 \log 0.08$

If logs are calculated to base 10, the entropy is -0.492; if to base 2, the entropy associated with the set is 1.63 bits.

Entropy so calculated has several important properties: (1) For any given set of probabilities, it is at a maximum when all the probabilities are equal; (2) different H's, derived from different sets, can be combined to yield an average entropy. Such combinations can be used to find the entropy appropriate to a Markov chain. In the illustration of the insect population, each column has a set of probabilities that total 1, and each therefore can provide an entropy.

Shannon defines the entropy (of one step of the chain) as the average of these entropies, each being weighted by the proportion in which that state, corresponding to the column, occurs *when the sequence has settled to its equilibrium.* . . . . Thus the transition probabilities of that section, with corresponding entropies and equilibrial proportions shown below are,

	B	W	P
B	1/4	3/4	1/8
W	3/4	0	3/4
P	0	1/4	1/8

Entropy: 0.811 0.811 1.061
Equilibrial proportion: 0.449 0.429 0.122

Then the average entropy (per step in the sequence) is 0.449 × 0.811 + 0.429 × 0.811 + 0.122 × 0.061 = 0.842 bits.[22]

The information given above about the insect population and its changes might have to be transmitted over a channel. The carrying capacity of the channel must be at least equal in entropy to the entropy of the information, per unit of time, for the channel to convey the information. One of Shannon's fundamental theorems states that for information of a given entropy per unit of time, a channel having this or greater entropy can carry the information and that a coding always exists by which the channel can be so used.

Much of Shannon's work is devoted to showing the possibilities of such relations and to demonstrating the devising of suitable codes. The "content" or "meaning" of information is of no concern to communication theory as developed by Shannon, only its variety and the means and probabilities of coding, channeling, and decoding information, in the sense given here, are pertinent.

Noise

If the output of a transducer can be reduced to two components, one a source of variety (information) and the other a nuisance regarded as "noise," information theory must provide for the latter's effects. Noise is not distinguishable from any other source of variety. Noise is a problem only when it interacts with a wanted message with some mutual destruction. Shannon has developed a measure for the degree of corruption of a message by noise, for incessantly transmitting channels. This involves knowing the probabilities of all possible outcomes: (1) No symbol sent—no symbol received; (2) no symbol sent—symbol received; (3) symbol sent—symbol not received; (4) symbol sent—symbol received. Once the long-term probabilities are known, a measure of *equivocation* can be devised, expressed in terms of bits per symbol. Shannon has shown that if the channel capacity is increased by an amount not less than E (equivocation), the messages can be encoded so that the fraction of errors still persisting may be brought as close to zero as is desired.[23]

Having thus established the conceptual bases of cybernetics, Ashby moves on to consider how regulation and control operate in complex biological "systems" and, by extension, in social and economic systems.[24] These are the central themes of cybernetics.

Regulation in biological systems is described as the development of mechanisms that ensure the survival of the organism. Forms that were defective in their powers of survival have been eliminated in the course of the earth's history, and the features of every life form bear the marks of being adapted to ensure that survival. The brain too is a means of survival. What survives over the ages is not the individual organism, but rather "certain peculiarly well compounded gene-patterns, particularly those that lead to the production of an individual that carries the gene-pattern well protected within itself, and that, within the span of one generation, can look after itself."[25]

The mechanism of survival operates in system terms:

$$\boxed{\text{D}} \rightarrow \boxed{\text{F}} \rightarrow \boxed{\text{E}}$$

D represents the source of disturbance and dangers from the world; E represents the set of essential variables that must be kept within specific limits for the organism to survive; F represents interpolated parts formed by the gene-pattern for the protection of E, which might include a shell, a brain, claws, and so on as well as parts of the environment, such as a rabbit burrow, a sword, armor. The set E of essential variables may be divided into two subsets—one corresponding to "organism living," or "good," and the other corresponding to "organism not living," or "bad."

Many devices illustrate these principles. A thermostatically controlled water bath may be designed to maintain water temperature within specified limits (E); D is the set of all disturbances—cold air, addition of cold water to the bath, and so on; F is the regulatory machinery, which by its action tends to limit the effects of D on E. Similarly, an automatic pilot is designed to maintain the position of an aircraft within set limits against gusts of wind, movements of passengers, irregularities of the thrusts of the engines, and so on.

The purpose of the regulator is to limit the flow of variety from the environment to the essential variables; an effective regulator in a sense prevents the essential variables from "knowing" that disturbances have occurred. The range of defense is quite wide, running from the interposition of purely passive blocks to disturbance—for example, the tortoise's shell or the human skull—to the opposite extreme of skilled, complex, and mobile defense. The fencer in a deadly duel wears no armor, and trusts to his skill in parrying.

Most higher organisms have developed a nervous system for defenses that are closer to the second type than to the first.

Cybernetics seeks to answer the following questions about the skilled active type of defense: What principles must govern it? What mechanisms can achieve it? What is to be done when the regulation is very difficult? [26]

Ashby treats this problem in terms of a formal game between two "players," D (disturbance) and R (regulator), symbolized by a table:

R

	α	β	γ
1	b	a	c
D 2	a	c	b
3	c	b	a

D must play first, by selecting a number, and thus a particular row. R, knowing this number, then selects a Greek letter, and thus a particular column. The italic letter specified by the intersection of the row and the column is the *outcome*. If it is an a, R wins; if not, R loses.

Examination of the table soon shows that with this particular table R can win always. Whatever value D selects first, R can always select a Greek letter that will give the desired outcome. . . . In fact, if R acts according to the transformation

$$\downarrow \begin{matrix} 1 & 2 & 3 \\ \alpha & \beta & \gamma \end{matrix}$$

then he can always force the outcome to be a.[27]

The table above is, of course, peculiarly favorable to R; other more complex tables can be developed in which R's chances of winning are less favorable, according to which targets within the table (a, b, c, etc.) are specified.

The relations between D, R, and the sets of possible outcomes can be formalized and mathematicized. Assuming that R selects a single-valued transformations, each transformation uniquely specifies a set of outcomes, and it can be seen that only variety in R can match or reduce variety in D. This is the Law of Requisite Variety: Only variety can destroy variety.

Ashby indicates that this law is of general applicability and is not a trivial outcome of the tabular form. He cites Shannon's work on variety spread out in time and of incessant fluctuation, demonstrating algebraically that the entropy of an outcome can be forced down below that of D only by an equal increase in the entropy of R. Furthermore, since this relation is mathematically proven, it is independent of any particular machine; it cannot be overthrown by the invention of some new device or electronic circuit. The law owes nothing to experiment.

Regulation as the set of responses by which an organism maintains its essential variables within acceptable limits against disturbances coming from the environment is fundamentally related to the game as described above. Furthermore, the table itself (T) represents those external or internal limitations R must take for granted. If R is a well-made regulator, R is a transformation of D such that all outcomes fall within the subset of acceptable outcomes. In this case T is also a factor in the set of relations, and R and T together act as the barrier to D.

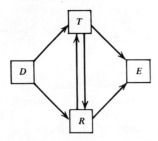

The arrows represent channels of communication. The variety in D determines the variety in R, and the variety in T is determined by the variety of both D and R. These may be embodied in actual machines and organisms in many ways. Thus, R's capacity to act as a regulator is its capacity as a channel of communication.[28]

Ashby's extension of cybernetic considerations to biological and social matters is illustrated by some problems he gives:

A certain insect has an optic nerve of a hundred fibres, each of which can carry twenty bits per second; is this sufficient to enable it to defend itself against ten distinct dangers, each of which may, or may not, independently, be present in each second?

A general is opposed by an army of ten divisions, each of which may maneuver with a variety of 10^6 bits in each day. His intelligence comes through 10 signallers, each of whom can transmit 60 letters per minute for 8 hours in each day, in a code that transmits 2 bits per letter. Is his intelligence-channel sufficient for him to be able to achieve complete regulation? . . . The general can dictate orders at 500 bits/minute for 12 hours/day. If his intelligence were complete, would this verbal channel be sufficient for complete regulation?[29]

Regulation and control suggest that the terms *constant* and *varying* are related and not necessarily contradictory. Thus a driver who steers a car accurately down a winding road may be regarded both as causing the steering wheel to show great change and activity and as maintaining the distance between the car and the edge of the road at a constant value. Much activity, movement, and change, insofar as it is coordinated or homeostatic, also points to other invariants and constancies.

THE ERROR-CONTROLLED REGULATOR

The problem of regulation is forming the mechanism R, given T, D, and the essential variables (E) that must be kept within specified limits. In some forms D is transmitted directly to R:

$$D \rightarrow T \rightarrow E$$
$$\uparrow \quad \lrcorner$$
$$R \leftarrow$$

In others, R derives its information from T

$$D \rightarrow T \rightarrow E$$
$$\uparrow\downarrow$$
$$R$$

Again, in others, R is affected only by the actual effect at E:

$$D \rightarrow T \rightarrow E$$
$$R \leftarrow \lrcorner$$

This is the basic form of the simple error-controlled servomechanism. "This form is of the greatest importance and widest applicability."[30]

An error-controlled regulator cannot be perfect, in the sense of successfully controlling all outcomes. Since R receives necessary information through E, it cannot eliminate all variety in E; in other words, the more successful R is in keeping E constant, the more it blocks the channel by which it receives necessary information. Thus error-controlled regulation can be only partial.

But in many situations it is sufficient, especially where essential variables lie along a scale of desirability—undesirability that is continuous, without a sharp demarcation of critical from nonessential values. Thus small errors are allowed to occur that by feeding information to R, permit regulation against large errors.

The schematic representation of the problems involved in regulatory devices is more general than may appear at first glance. Thus compound disturbances can still be represented symbolically as D, simply by making D a vector with any number of components. Again, suppose that T is "noisy." If T were an electrical machine whose action is affected by variations in voltage, this could be managed simply by redrawing the boundaries between what is represented as T and what is represented as D. D, T, and E have been defined, Ashby reminds us, in purely functional form, and D represents "that which disturbs." This redefinition of boundaries is not purely symbolic or magical; this new transformation may allow regulation to be restored, no doubt in more complex form. Similarly, if the essential variables in E that must be kept within specified limits are complex, conditional, and interrelated, E also can be redefined as a vector having any number of functional components. The same can be said for all other details in the scheme.

Thus the representation of regulation in a rectangular table played by two opponents, D and R, is much more general than it seems and can include cases of any degree of internal complexity.[31]

MARKOVIAN REGULATION

Regulation of industrial processes by servomechanisms usually employs determinate machines or systems. Markovian machines, governed by probabilistic states, operate by procedures that seem much more fumbling and casual when compared with the smooth regulation of servo-mechanisms. "Nevertheless, living organisms use this more general method freely, for a machine that uses it is, on the whole, much more easily constructed and maintained, for the same reason it tends to be less upset by minor injuries. It is in fact often used for many simple regulations where speed and efficiency are not of importance."[32] The sole difference between a Markovian machine and a determinate one is that in the former its trajectory is not unique; but no difference in principle exists. Markovian machines, however, may appear to be different in that their behavior is often a path through a maze or something similar. A Markovian machine may appear to operate by trial and error, though Ashby finds the term unfortunate. There are many trials aimed at one solution. He prefers to describe the operation of such machines as "hunt and stick." Markovian machines appear to operate by the process of hunt and stick, searching apparently at random for what they want and retaining the desired states when reaching them. As long as the long-term probabilities of Markovian machines are stable, they exhibit forms of determinate regulation.

GAMES AND STRATEGIES

Ashby considers von Neumann and Morgenstern's theory of games to be essentially related to regulation and control in biological organisms:

> . . . the inborn characteristics of living organisms are simply the strategies that have been found satisfactory over centuries of competition, and built into the young animal so as to be ready for use at the first demand. Just as many players may have found "P-Q4" a good way of opening the game of Chess, so have many species found "Grow teeth" to be a good way of opening the Battle of Life.[33]

Ashby regards his basic formulation of regulation and control to be identical with the payoff matrix that is fundamental to the theory of games, which is in fact isomorphic with that of certain machines with input. Related to this is the theory of military codings and decodings as exemplified in Shannon's *communication theory of secrecy systems*. Advances in one field usually throw light on the others.

The next step is that of considering regulation in very large systems—brains and societies.

Ashby assumes the applicability of cybernetic thinking to social matters. Ecologists, he says, may want to control epidemics, economists to control economic slumps, psychotherapists to regulate the functioning of sick brains, and he adds, "the sociologist faces a similar situation."[34]
Largeness of the system is not the problem; rather, it is the variety in the disturbances that must be regulated against. Regulation may sometimes be achieved by redefining what is acceptable—that is, standards may be lowered. Another possibility is that of increasing the scope and power of R until its capacity is adequate. Disturbances may be repetitive—for example, variations in temperature—or may be rare and nonrepetitive—for example, the appearance of a supernova. The latter type of disturbances cannot be planned for fully because they show no constraints. Regulation is mainly focused on repetitive disturbances. If regulation is the responsibility of some entity (Ω, Ω has the option of either *being* the regulator or of building a machine that

will act as *R* and will carry out regulation of indefinite length without further action by Ω. We are thinking here not so much of the engineer at his bench as of the brain that wants to develop regulatory machinery in the nervous material available to it within which it will implant learned reactions. With the formula Entity Ω designs machine *M*, we include cases such as (1) genes determining formation of the heart, (2) a mechanic making a bicycle, (3) a part of the brain determining the internal connections in a nerve net, (3) a manager laying out a factory to determine production along certain lines, (5) a mathematician programming an automatic computer to behave in a certain way.

In all of these cases *"the act of 'designing' or 'making' a machine is essentially an act of communication from Maker to Made,* and the principles of communication theory apply to it."[35] The designer's act of selecting one model from a set of many possible models is equivalent to a determining factor that fixes an input at a permanent value. Thus a part of a brain may behave towards another part as designer to machine, and one machine can design another. Design means simply selecting a subset of possibilities from a larger set, and a machine may do this. The ultimate problem is selecting the regulator that chooses among the possibilities. How is a regulator to be made? By another regulator. Is this a *reductio ad absurdum?* Not necessarily, for when a regulator R_1 acts so as to bring into existence a regulator R_2, the capacity of R_2 is not bounded by that of R_1. The fact that the earth has existed a long time, with selection acting throughout this time, accounts for the presence on earth today of many good regulators. No further explanation is necessary.

Again, what does this imply when the very large system to be regulated is the social and economic world and the responsible entity Ω is some set, of sociologists, perhaps whose capacity, as a regulator, is limited to that available to the members of the species *Homo?*[36]

The very fact demonstrated by biological history, that regulators of greater capacity have been produced by regulators of more limited capacity, suggests that the same can happen in the sphere of intelligence so that an *amplification* occurs.

Amplification devices are common in engineering of all kinds—a few pounds of pressure exerted by a hand or arm on a lever is converted to many hundreds or thousands of pounds of pressure exerted by a mechanical device. For the amplification to occur, energy is drawn from stores in the environment (coal, steam, electrical power, etc.). The question for the cyberneticist is: Can intelligence be amplified by similar procedures? Since problem solving (which is what intelligence is) is a matter of selection—that is, of selecting an appropriate answer or response to a given problem—the problem is formally no different from selection processes of any other type. What we speak of as "high intelligence" is really only "power of appropriate selection," he says. A talking black box that showed such a power would have to be called intelligent. If physical power can be amplified, so can intellectual power be amplified. Gene patterns do it every time they form a brain. What is new, he tells us, is that now we can do this consciously. But, he concludes, "these are not matters for an Introduction." [37]

This arcane knowledge, we must suppose, will remain esoteric. Ashby's other major statement on cybernetics is found in his *Design for a Brain—The Origin of Adaptive Behavior*. [38] It places less emphasis on sociological matters, however, and so may be described in less detail. The central problem here is to account for the origin of the nervous system's unique ability to produce adaptive behavior, on the hypothesis that the nervous system is essentially mechanistic. Ashby seeks to develop a logic of mechanism in mathematical form. He suggests that the brain uses methods that have been used in machines either little or not at all. An explanation of how the brain operates in this respect could explain the adaptiveness of man. The principal concern is not with reflex action but with learned behavior. The problem is one of self-coordination—each part's working (in the system) must be related to all the other's before it can be evaluated as correct or incorrect.

Ashby sees the brain as an organ developed in evolution as a specialized weapon of survival; but the nervous system and living matter in general is assumed to be essentially similar to all other matter. Therefore, no use of any "vital" property or tendency is claimed. "The sole reason admitted for the behavior of any part will be of the form that its own state and the condition of its immediate surroundings led, in accordance with the usual laws of matter, to the observed behavior." [39]

His approach is determinist, operational, and nonteleological. Psychological concepts are avoided unless they can be shown in objective form in nonliving systems. "It will be assumed throughout that a machine or an animal behaved in a certain way at a certain moment because its physical and chemical nature at that moment allowed it no other action. Never will we use the explanation that the action is performed because it will later be advantageous to the animal," because the purpose is to explain the origin of behavior that appears to be teleologically directed.

Ashby further states the assumption that the functioning units of the nervous system (and of the environment) behave in a determinate way. He does acknowledge that the ultimate units—the atoms—are essentially indeterminate; but since the units he deals with, such as the neuron, the relay, a flow of current, are much larger and we deal with *average* properties of many

atoms, they are assumed to be determinate. Thus his problem is (1) to identify the nature of the change that shows as learning, and (2) to find why such
changes should tend to cause better adaptation for the whole organism.

The principal formal device by which he accounts for adaptation is that of
step-function. We must distinguish changes that constitute behavior from
changes that constitute shifts from one behavior to another, where a line of
behavior is conceived of as a fixed determinate "trajectory." The basic operation by which the experimenter is to win new knowledge from the "machine"
is to use his power of control to determine particular states in the system and
in the environment of the system. He then observes and records the system's
transitions.[40]

Of course he may also observe chains of transitions. Thus a system can be
followed along a *line of behavior*. Results can be represented geometrically in a
"phase-space." If a system is released repeatedly from all possible initial
states within a particular set of surrounding conditions, all the resulting lines
of behavior comprise the system's *field*, which defines the characteristic behavior of the system. A "machine with input," then, is one whose behavior is
always the same under a given set of conditions—a "state-determined system."

Ashby explicitly bases his work on the assumption that a complete knowledge of the present state of a system furnishes sufficient data to determine
definitely its state at any future time or its response to any external influence.

Laplace's assumption of determinacy of the entire universe is now known
to be false at the atomic level. But, says Ashby, since the assumption appears
to hold true over "great ranges of macroscopic science," he adopts it extensively.[41]

Ashby explicitly relates cybernetic thinking to Newton and Darwin and to
a determinist conception of the universe. He further assumes that the living
organism is no different in its nature and processes from other matters. The
truth of the assumption is not discussed; he deals instead only with the technique of applying this assumption to the complexities of biological systems.
Further, he assumes that all organismic qualities can be either numericized
or ignored.

STABILITY AND ULTRASTABILITY

In terms of survival value, stability refers to the maintenance by a system of
essential variables within acceptable limits. It applies not to a material body
but to a field; thus a stable region is one whose lines of behavior all remain
within the region. Stable fields include systems such as a thermostat that
maintains temperature within a steady oscillation. Stability is a property of a

whole system and cannot be assigned to any one part of it. Adaptation in biological systems is seen as behavior that maintains stability. In this respect Cannon's work on homeostasis was a major pioneering achievement. The mechanisms of adaptation may operate partly through the organism and partly through its environment. The problem, again, is to account for the way in which a determinate machine changes from a form producing chaotic unadapted behavior to a form in which parts are so coordinated that the whole is stable.

Here the step function plays its central role. Step functions are a set of regulatory functions that operate by discontinuities, in contrast to continuous functions. A step function shows constant values except for discontinuous jumps to other levels. Thus any variable acting on an "all-or-none" basis behaves as a step mechanism.

The role of the step function in the total system is to provide the possibilities of adaptive behavior. Thus a double feedback system exists between regulator and environment and between the essential variables and the regulator. The first feedback consists of ordinary sensory inputs; the second goes through the essential variables and indicates whether they are within essential limits. The first feedback plays its part *within* each reaction; the second determines which reaction shall occur.

Thus organisms are always mechanisms whose behavior is at all times determined (state-determined); but since many vast combinations of step functions may exist, such responses will appear to us to be spontaneous. The purpose of the step function is to change value only when essential variables pass outside given limits and so switch the organism to a different mode of reactive behavior. Ashby evidently believes that this scheme successfully eliminates any reference to teleological concepts.

For the rest, he devotes himself to showing the ability of cybernetic terminology and schemas to parallel features of biological systems such as adaptability, stability, and the ability to learn by trial and error. He does this, however, by assuming that stability is inherent in the nature of systems and of natural processes of change. He illustrates this by the construction of an electrical device, the homeostat, which is subject to "disturbances" induced by the experimenter and which generally exhibits stability when the disturbances are kept within limits.

THE HOMEOSTAT

Ashby uses the homeostat as a central image. It consists of four units, each of which carries a magnet on a pivot. Each unit emits an electrical output proportional to the deviation of its magnet from a central position. Thus each unit receives current at a constant level; and if just this level is maintained, the magnet remains at rest. But if this level is either exceeded or fallen short of, the difference in current in either direction is sent to the magnet, which then deviates either to the right or the left. Next the units are joined together such that each sends its output to the other three and thereby also receives inputs from the other three.[42]

Step functions are provided by a set of four switches, one for each unit. Each switch has 25 positions; thus more than 390,000 combinations of parameter values are available; these switches play the part of step functions for each unit, causing discrete changes in levels of current supplied to each unit. The specific parameter values to be provided to the switches are determined by using a table of random numbers. Thus if any magnet diverges from its central limits for more than a few seconds, the values of the switches are transferred to other random values.

The experimenter could induce displacements by a variety of ways and then examine the behavior of the system as a whole. If a particular disturbance destroyed stability, uniselector (switch) changes would occur until stability returned. The homeostat shows the power of self-reorganization in elementary form.

Ashby reports how the homeostat parallels learning processes in laboratory animals: Using three of the four units, he wished to induce the homeostat to respond to a forced movement of unit 1 with an opposite movement of unit 2; if it failed to do so, unit 3 would be forced over to an extreme position (the fourth unit was disconnected for this experiment). At first try unit 1 was moved, and unit 2 moved similarly; this was the "forbidden" response, so unit 3 was forced to an extreme position, and the step mechanisms changed value. This was repeated and again the forbidden response occurred; again unit 3 was forced to an extreme position. At the third try the "response" was the desired one, so no further deviation was forced on unit 3. At further trials, no more deviations occurred. Thus the line of behavior at which units 1 and 2 started was unstable. Ashby indicates that in psychological terms to say that the "trainer" has "punished" the "animal" is equivalent to saying in cybernetic terms that the system had a set of parameter values that made it unstable.

Ultrastable systems can adapt to two environments, such as the diving bird that adapts to water and to land or the kitten that adapts both to the mouse that flees into a hole or a bird that flies upward. The analog in the homeostat is a commutator switch that can reverse polarity—that is, the equivalent to adaptation to two environments.

If presented with an "unusual" environment, the ultrastable system produces a set of step-mechanism values that results in a stable system. Thus if vital variables show "unusual" values, the step mechanisms exhibit compensatingly unusual values.

Thus when some "unusual" conditions were imposed on the homeostat, the resulting instability induced changes in the step mechanisms until stability was found.

There are, of course, unusual problems for which the homeostat's repertoire contains no solution and could result in shorting a relay; in such a situation the homeostat fails to adapt; but so would a living organism fail to adapt to a problem if its total repertoire contained no solution for it.

Although the higher animals show a richness in adaptations that appears to transcend the crude model of the homeostat, Ashby argues in detail that the living brain adapts by ultrastability. Although an ultrastable system may fail to adapt to an environment offering sudden discontinuities, so too will a living organism; in this way, for example, traps catch animals. But the general stability and continuity of natural environments on earth has made adaptation by these means possible and in fact inevitable.

As noted before, Ashby regards adaptation and stability as "lines of behavior" or as "processes" that "operate" not merely within organisms, but through total systems that include organisms and their environments. Thus, he says, the finer points of catching mice are taught to the kitten by the mice themselves. Ashby argues repeatedly that lines of behavior tend to converge toward stability just by the general stability of the earth's environment. But then Ashby appears to arrive, evidently without noticing that he does so, at a hypostatization of "stabilization processes" or "adaptation processes" that operate both through organisms and their environments; he hopes to escape a teleological approach by ascribing stability-seeking lines of behavior to the world in general. He appears, therefore, to have switched to a point of view in which stabilizing lines of behavior become the ultimate reality.

The difficulties inherent in his line of reasoning should now perhaps be made evident. The entire theory of cybernetic mechanism is dependent on the construction of logically closed determinate systems; once granted, the formal and mathematical bases of many types of "feedback," "stability," and "adaptation" can be developed in mathematically rigorous detail. But at crucial points in his presentations Ashby makes the retreat (or leap) from the empirical world into a selected subset of logically closed possibilities that he then manipulates and expands to the point where the closed system eclipses and replaces the empirical world—and ultimately stands in its place.

NORBERT WIENER

Wiener argues that cybernetics will have an immense impact on civilization, not only in the form of the automation of a great variety of functions previously performed by humans, but perhaps even more in its philosophical import. Wiener argued that cybernetics in its full development had major implications even for theology, affecting our conceptions of both God and man.

We are not concerned here primarily with the nature or the technicalities of cybernetics itself, but rather with its social and philosophical claims. Nevertheless, some of the intrinsic content of cybernetics is important in that its imagery and terminology have been adopted by system theorists. Like W. Ross Ashby, Wiener displays considerable sympathy for the extension of cybernetic concepts in other fields. Cybernetics, furthermore, appears to shade off into communications theory; in some respects both Wiener and Ashby regard them as virtually identical in concerns, techniques, and mathematical symbolism.

In his first general *summa* in the field, *Cybernetics*, Wiener offered both a mathematical basis for cybernetics and drew some of its implications for society. *The Human Use of Human Beings—Cybernetics and Society* was then written as a nontechnical nonmathematical revision of the earlier work in which the importance of cybernetics for society was more explicitly stated. *God and Golem, Inc.*, represents Wiener's theology.

It is the thesis of this book that society can only be understood through a study of the messages and the communication facilities which belong to it; and that in the future development of these messages and communication facilities, messages between man and machines, between machines and man, and between machine and machine, are destined to play an ever-increasing part.[43]

Wiener begins by noting that physical science has replaced the image of a determinist universe with that of a contingent one, by way of the physics of Boltzmann and Gibbs, who introduced the statistical and probabilistic treatment of particles into physics. He in fact indicates that his book is devoted to presenting the impact of Gibbs' point of view on modern life: In Gibbs' universe order is the least probable, chaos the most. Although the universe tends to increasing entropy (disorder), there are, as it were, islands or enclaves within which organization tends to increase. The tendency is, no doubt, limited and temporary. But among these islands life is at home. This is the starting point of cybernetics.[44]

This new science had its forerunners in earlier concepts of the machine. The earliest machines were, of course, clockwork devices that operated blindly. These were followed by elementary machines with "sense organs"—that is, receptors for messages. A simple illustration is the automatic door that is activated by a photoelectric cell; another is the thermocoupling device in thermostats, in which changes of temperature affect electrical conductivity and so activate heaters to control the temperature. More complex forms are exemplified by machines such as antiaircraft guns that possess sensing devices (usually radar), built-in instructions for computing the trajectory of the aircraft sensed and for calculating the path of a shell to be fired, not at the sensed position of the aircraft, but at its position computed to be the point of arrival of the shell. The structure of such devices—for example, in modern computers is: (1) data input, (2) memory, (3) instructions ("taping"), (4) output. The essential feature is that the machine must operate according to *feedback*—the control of a machine on the basis of its actual performance. These mechanisms produce local and temporary reversals of entropy.[45] In fact we know that human muscular reactions are controlled and regulated in precisely the same way: The reflex by which one reaches for an object is modified by information fed back to the brain about the distance between the object and the hand that reaches for it. Wiener takes for granted the significance of feedback not merely for muscular or mechanical control, but for social control as well. Not only is there a direct parallel between the physical functioning of living individuals and the performance of some of the newer machines; not only do they take in information about their performance, which becomes available for later stages of the performance; this operates on the societal level. More attention should be paid to this.

This complex of behavior is ignored by the average man, and in particular does not play the role it should in our habitual analysis of society; for just as individual physical responses may be seen from this point of view, so may the organic responses of society itself. I do not mean that the sociologist is unaware of the existence and complex nature of communications in society, but until recently he has tended to overlook the extent to which they are the cement which binds its fabric together.[46]

Thus, claims Wiener, the theory of messages is of philosophical significance in the striving of messages to hold back nature's tendency toward disorder. This significance relates to various earlier views of the universe. The paradigm of the concept of entropy is the second law of thermodynamics: the tendency for entropy to increase. But of course in systems not completely in equilibrium, entropy does not increase; locally, it may even decrease—a phenomenon that is not only local but temporary. In the universe we find enclaves of increasing organization. These are not merely living beings. Machines also contribute to the local and temporary buildup of information.

There is no reason, says Wiener, that we cannot regard these machines as resembling human beings in their ability to reverse entropy.[47]

The nervous system and the automatic machine are fundamentally alike in that they make decisions on the basis of past decisions. Whether a machine will decide between two alternatives, such as the opening or closing of a switch, or a nerve fiber will decide to carry an impulse or not, the analogy is close and detailed.

The machine, like the organism, is a device for working against entropy. The scientist working to discover the order and organization of the universe is in effect playing a game against the arch enemy, disorganization. This raises the important question whether this enemy is "Manichaean or Augustinian." If Manichaean, the devil of disorganization is active and creative and uses any trick of craftiness or dissimulation to confuse his opponent, the scientist. In St. Augustine's view evil was not an active creative principle, but rather a matter of incompleteness, narrowness, limitation. The Augustinian devil is not a power in itself, but simply a measure of human weakness and limitation.

Wiener decides that the scientists' opponent is an Augustinian, not a Manichean, devil. "Nature offers resistance to decoding, but it does not show ingenuity in finding new and undecipherable methods for jamming our communication with the outer world."[48]

Thus the scientist is disposed to regard his enemy as honorable, an attitude necessary for his effectiveness as a scientist, but one that handicaps him because it makes him the "dupe of unprincipled people in war and politics." It also makes him incomprehensible to the general public, which is concerned less with nature as an antagonist than with personal opponents. Wiener is not free of professional self-righteousness.

He then addresses himself against the notion of progress. "There are local and temporary islands of decreasing entropy in a world in which the entropy as a whole tends to increase, and the existence of these islands enables some of us to assert the existence of progress." The Enlightenment gave us the notion of progress, and the line of thought from Malthus to Darwin was destined to overthrow the notion. Evolution produces chance variation, and the survivors are more adapted to the environment than those that do not survive. These residual patterns of forms of life give the appearance of universal purposiveness. W. Ross Ashby finds similar processes among machines; even an arbitrarily constructed machine tends to seek out a purpose; it tends to favor modes of activity in which the parts work together in a stable way.

I believe that Ashby's brilliant idea of the unpurposeful random mechanism which seeks for its own purpose through a process of learning is not only one of the great philosophical contributions to the present day, but will lead to highly useful technical

developments in the task of automization. Not only can we build purpose into machines, but in an overwhelming majority of cases a machine designed to avoid certain
pitfalls of breakdown will look for purposes which it can fulfill.[50]

Thus Darwin makes the first breach in the notion of progress. Marx and
his contemporary socialists, Wiener says, accepted a Darwinian point of view
in the matter of evolution and progress. It is hard to know what Wiener
means by this, inasmuch as Marx's view of history could be described as
optimistic in the sense that history was destined to end in universal social
harmony. Second, Ashby's "brilliant idea" is not at all original, but is in a
sense only a copy of Darwin's view of the mechanism of evolution. Furthermore, he argues against the notion of progress as if he were the first and only
one to claim that this emperor has no clothes. He seems to feel that the notion
of progress is unshakable in the popular and business mind of America, which
as soon as it hears of the second law of thermodynamics, will begin to think
otherwise. But, he says, although we are shipwrecked passengers on a doomed
planet, at least we can go down in a manner befitting our dignity.

Since cybernetics is a theory of communication, it must perforce examine
the nature of language. Human communication is distinguished from animal
communication by the delicacy and complexity of the codes and the high
degree of arbitrariness of the codes used.[51]

Animals, says Wiener, convey to one another crude emotions of fear, warning, and the like. "In general, one would expect the language of animals to
convey emotions first, things next, and the more complicated relations of
things not at all."[51] Their languages, furthermore, are fixed and unchanging.

Wiener further notes that machines can be created that have linguistic
capacity.

His discussion leads him to a major problem in communications theory as
he conceives it: the fact that messages can be garbled or distorted in transmission, whether in the speech of two persons or in the form of messages transmitted over electric lines. Transmission is never perfect; something is always
lost along the way. "That information may be dissipated but not gained, is, as
we have seen, the cybernetic form of the second law of thermodynamics."[53]

Ordinary language systems terminate in the "special sort of machine
known as a human being." The human being seen as a terminal machine has
a communication network that can be considered at three levels. The first
level is the human ear, plus biological mechanisms that transmit sound to the
brain. The second level is semantics. Here Wiener speculates on the neural
mechanisms in the brain that translate sound into sense. The third level is
behavior, both in terms of actions in meaningful response to linguisitic communications and in the form of responsive speech. Animals such as chimpan-
zees, no matter how in infancy they may outstrip the performance of human

infants, have no built-in mechanisms for translating sounds into meaning systems. Speech is the greatest interest and most distinctive achievement of man; primitive man has always stood in awe of language and has surrounded names with a sense of magic. But the ultimate significance of communication is that it serves to bind societies together. In fact modern communication makes possible—and even inevitable—the establishment of "the World State."[54]

But the communications practitioner must struggle against entropy, the tendency for information to leak in transit unless external agents are introduced to control it. There are, of course, two types of communication: that animated by the desire to convey information and that found in law, politics, and other contentious forms of communication animated to impose a point of view against a willful opposition. An adequate theory of language seen as a game must distinguish between these two approaches. But from the point of view of cybernetics, semantics is to be understood as defining the extent of meaning in a system and controlling its loss within that communications system.

Wiener offers an analogy that he considers to be of central importance: The organism may be considered as analogous to a message. The organism is a pattern that maintains itself against chaos and disintegration; the message is a pattern that imposes itself upon the chaos of "noise."

No organism is an identity in a material sense; studies of metabolism show that the material elements of a body are exchanged with the environment at a rapid rate so that a living organism may in the course of its existence be composed of entirely "new" material elements many times in succession. The organism is not matter, but a pattern that maintains itself, a whirlpool in the stream of matter, which may be likened to a flame whose identity is not in its elements but in its form.

Exactly the same thing may be said of a message. Theoretically there is no difference between a message and the pattern we call an organism. In fact an organism could be "transmitted" over appropriate transmitting apparatuses and reconstituted somewhere else by an appropriate receiving instrument. There is no fundamental distinction between material transportation and message transportation.[55]

The implication for the modern world is evident. As the network of communications extends itself ever more efficiently over the earth, world society can knit itself much more closely together. The communication of a man's message is an extension of himself over the earth.

The notion of patterns (whether organisms or messages) that can be transmitted implies that such patterns can be duplicated. This of course raises the question of the existence of the soul. The Christian view is that the soul is created at the moment of conception and continues in some spiritual form even after the end of earthly life. Buddhism also considers the soul as continuously existing, although after leaving one form of organic life it passes into another; thus the soul of a man may enter the body of an animal. Wiener refers to Leibniz' philosophy of monads, inexplicably describing it as "the most interesting early scientific account of the continuity of the soul."[56]

But, says Wiener, the modern view is that the continuity of existence for an organism may terminate even without death. Cell division produces *twinning*—where formerly there was one cell, now there are two genetically identical cells. The phenomenon also manifests itself in psychology in what is known as the split personality. On the basis of such phenomena, we can assume that the pattern of an individual organism can be duplicated in the same way that the pattern of a machine can be. But the message—the amount of information necessary—may be extremely complex. There is, however, no theoretical problem, only a technical one. Furthermore, for most practical purposes, the traffic in bodies is less important than the traffic of information.

Wiener believes that these considerations lead logically to the problem of law and communication. He defines law as the ethical control of communication. Such control involves a conception of justice and techniques for making these concepts effective.

Concepts of justice do vary from one society to another, but Wiener feels that he can offer a definition of justice. It is summarized in the famous slogan of the French Revolution: *Liberté, Egalité, Fraternité.* Wiener understands *liberté* to mean the right of each individual to develop his full potential; *egalité* is the feeling that what is just for one is just for another; *fraternité* means the sentiment of unlimited good will. Whatever compulsion the state or the community exerts to ensure their existence must be minimal. But of course developing a fair legal code is difficult. Legal responses to conflicts and lawsuits should have the following properties: Contracts and decisions should be unambiguous, free from duress, and consistent. In a word, legal problems are cybernetic problems.

Let us put it this way: the first duty of the law, whatever the second and third ones are, is to know what it wants. The first duty of the legislator or the judge is to make clear, unambiguous statements, which not only experts, but the common man of the times will interpret in one way and in one way only. The technique of the interpretation of past judgments must be such that a lawyer should know, not only what a court has said, but even with high probability what the court is going to say. Thus the problems of law may be considered as communicative and cybernetic—that is, they are problems of orderly and repeatable control of certain critical situations.

There are vast fields of law where there is no satisfactory semantic agreement between what the law intends to say, and the actual situation that it contemplates.[57]

Unfortunately, the greatest obstacle to the contemplated cybernetic control of justice is the fact that legal discourse is a game, in the von Neumann sense, in which the conflicting parties seek not to uncover the truth, but to jam the messages of the opposing side—to bluff, to reduce opponents' statements to nonsense, and the like.[58]

Communications and cybernetics meet with obstacles beyond those of the law. At a time when the network of world communication is more efficient and complete than ever before in history, we see nations such as the United States and the Soviet Union adopting policies of secrecy that we have seen before only in the crafty policies of Venice in the days of its power. Ideas and scientific discoveries are treated like secrets of state; or, worse, they are treated like commodities to be patented, copyrighted, and marketed. The treasuring of original works of art, rare objects, first editions, and the like is on the same level. A high level of artistic taste can be developed by someone who has never seen an original, but only reproductions. All such barriers inhibit the free exchange of ideas and information and should be abandoned. In the field of science such secretive approaches coupled with the increasing bureaucratization of research can only, in Wiener's view, lead to disaster. Were it not for the narrow-mindedness of commercial, political, and organizational interests, the scientist would be much freer to do his work and to benefit the world by it. In fact his entire book, says Wiener, is based on the argument that the integrity of the channels of communication is essential to the welfare of society. But communication is debased and trivialized by the mass media; they produce standardized bland products like commercial loaves of white bread. Even in scholarly and scientific work the preference is for the familiar, the routine, and the standardized.

By this point it becomes evident that Wiener intends to apply the metaphor of cybernetics to a wide variety of fields and that he uses this as the occasion for sermons that repeat the claim of cybernetics to be fruitful for law, education, government, science, and so on. But that fails to engage in the traditional problems and dilemmas of these fields. Wiener describes what he calls the two industrial revolutions. The first was defined by the substitution of the machine for human muscle power; its great impact on society is likely to be dwarfed by that of the second industrial revolution, which we are facing now. This is symbolized by the appearance of the vacuum tube, the science of electronics that has emerged from it, culminating in the cybernetics concept of feedback, which when properly built into industrial processes, can result in computer governance of entire manufacturing processes (automation). By coupling feedback directly to the machine instead of through the agency of a human operator, immense new possibilities unfold. We may expect vast economic and social consequences to emerge from these developments, but the most he can tell us is that these new developments have great potential for good or for evil and that much depends on how society uses them.

Cybernetics can, Wiener continues, play important roles in medicine, especially in the area of prosthetic devices and in the analysis of certain diseases. One disease, characterized by intention tremor—the inability of the sufferer to guide his muscles accurately—appears to be a function of the breakdown of feedback mechanisms in the nervous system and brain. Parkinson's disease, characterized by tremors when at rest that disappear when action is undertaken is another form. Machines have been build that incorporate purposes—for example, to flee from or approach a light source—with feedback mechanisms that could be experimentally deranged. Much can be learned from these devices. Other machines that operate on behalf of the deaf or blind, attempt to provide sensory input in other forms: Speech may be translated into visible or tactile sensations for the benefit of the deaf or blind. These include the Bell Telephone Laboratories' "Vocoder," which translates voice patterns into visible images, and a "hearing glove" developed at M.I.T.

Of even greater import, however, are machines that have learned to play chess and checkers and to improve their games on the basis of their past experiences. These advances are of major societal importance, especially in their potential for developing and evaluating military tactics.[59]

Wiener cites an important review of his book *Cybernetics* by a Dominican friar, Père Dubarle, in the Paris journal *Le Monde* (28 December, 1948). He quotes Dubarle's review at some length, considering it of major importance. Dubarle is fascinated by the prospect of the rational conduct of human affairs, especially those that seem to exhibit statistical regularity, such as the development of public opinion.

The machine, given data on social and economic developments and an "average" human psychology, might measure the "most probable" outcomes. Of course human realities do not permit clear-cut determinations, only probable ones; this makes the task more difficult, but not impossible. The *machine à gouverner* may eventually supplement the inadequacy of the human brain in the political sphere; in the struggles for power that the theory of games studies.

The *machines à gouverner* will define the State as the best-informed player at each particular level; and the State is only the supreme co-ordinator of all practical decisions. There are enormous privileges. . . . They will permit the State under all circumstances to beat every player . . . by offering this dilemma: either immediate ruin or planned cooperation.[60]

Nevertheless, Father Dubarle remarks, the *machine à gouverner* will not be quite ready in the near future. The situation is and will remain too complex. There are only two conditions under which the social factors could be stabilized enough to meet the requirements of mathematical treatment: (1) sufficient ignorance on the part of the mass of the players, who will be exploited by a more skilled player, and (2) sufficient willingness on the part of the players to defer their decisions to one or two players who have arbitrary privileges.

This is a hard lesson of cold mathematics, but it throws a certain light on the adventure of our century: hesitation between an indefinite turbulence of human affairs and the rise of a prodigious Leviathan. In comparison with this, Hobbes' *Leviathan* was nothing but a pleasant joke. We are running the risk nowadays of a great World State, where deliberate and conscious primitive injustice may be the only possible condition for the statistical happiness of the masses; a world worse than hell for every clear mind.[61]

Wiener, in response to Dubarle's comments, seems to feel that the danger lies not in such machines, which are still too imperfect, but in the way they may be used by some humans to increase their control over the rest of the human race; or that political leaders may attempt to control their populations not by machines, but through techniques as narrow and indifferent to human possibility as if they had been conceived mechanically. But even without a machine for governing, there are groups that develop new concepts of war, economic conflict, and propaganda on the basis of von Neumann's theory of games.

Wiener assumes that the best safeguard against cybernetic tyranny would be the engagement of a philosopher and an anthropologist on the governing committee. Once we know what man's nature is and what his "built-in purposes" are, and once we know why we wish to control him, we can then wield this knowledge as "soldiers and as statesmen."[62]

Wiener appears to have missed the irony of Père Dubarle's remarks and to offer a cure that is an extension of the disease.

He concludes his somewhat shapeless book with a section on language and confusion. Communication is to be understood as a game played by speaker and listener against the forces of confusion. But human agents may try to jam communication. Nature does not do so. In this connection we must recall the famous saying of Einstein: "God may be subtle, but he isn't plain mean." Nature does not try to jam the message seeking of the scientist; she plays fair. If, after climbing over some obstacles, the physicist spies a new set of obstacles, he is sure that they were not put there just to thwart him. The scientist is justified in being naive in assuming that he is dealing with an honest God. Unfortunately, the professionally generated naivete of the scientist places him at a disadvantage in dealing with the political forces of the world, who are by nature more subtle, disingenuous, and "Manichean" in outlook than he is. He concludes with a plea for science as a way of life.[63]

Wiener's final work *God and Golem, Inc.* offers little more than has been said in his earlier works. He argues that cybernetics impinges upon society, ethics, and religion; he wishes to show how. First, some religious and scientific prejudices must be overcome. The former is the belief that man is essentially different from the animals; the latter is that living beings and machines are profoundly different. Once we overcome these taboos, which sterilize our thinking, we can proceed to gain new knowledge.

Cybernetics impinges upon religious issues in three respects: (1) There are now machines that learn (computers that improve their chess and checker games on the basis of past experiences); (2) there are machines that can make machines in their own image (they reproduce); (3) machines and living beings form symbiotic relationships. Thus the man who plays games with the machine he has constructed is analogous to God's struggle with man and with Satan.

But human learning must be considered under two aspects: (1) ontogenetic (the learning that an individual acquires in the course of his life experience) and (2) phylogenetic (the learning of the entire human race in the course of its evolution). He reiterates the concepts of Darwinian natural selection: (1) heredity (a species reproduces itself through heredity), (2) variation (chance variations arise in genetic combinations, giving rise to varieties of forms), (3) natural selection (some of these new forms prove less viable, others more). Thus the evolution appears to be moving in a given direction, but is actually a resultant of chance variation and natural selection. Central to this process is the fact that given the proper nutritive medium, genes create their own duplicates; similarly, man reproduces himself; and finally, machines can do exactly the same. Here, Wiener reiterates the image offered in his earlier books, of a black box of unknown internal construction, whose entire performance and nature can be recorded on a set of white boxes, (devices whose function and construction are known). In other words, giving a black box an appropriate receptive medium allows it to reproduce itself. In the course of his discussion Wiener indicates several affinities with the systems thinking of von Bertalanffy, especially with the notion that systems of one kind may give knowledge about other kinds of systems.[64]

For many, such speculative implications of cybernetics smacks of the old medieval fear of sorcery; such fears still exist, and in fact such a sin still exists. Among those who would use modern automation for profit or to release nuclear warfare, the applied engineer, the manufacturer, the politician are infected with the sin of sorcery. This spirit infects international relations through plans for war, especially in the form of games theories and computer-operated missile systems.

Thus, says Wiener, one of the great future problems we must face is how man and machines must interact and which functions properly belong to each.[65]

Machines work better and faster than man and are relatively tireless, but man is more complex and has a greater storage capacity for information. In addition, Wiener indicates that the human brain can also handle vague ideas, imperfectly defined, in the form of poems, novels, and paintings, which any computer would reject as formless material. These comments appear to be meant seriously.

The creative symbiosis of man and machine occurs at present in three areas: (1) prothestic devices, such as artificial hands activated by impulses coming from nerve endings in the arm; (2) machines for translating from one language into another, which although still too rudimentary for use in inter-governmental diplomacy, are improvable; (3) diagnosis of medical conditions in human beings. Here Wiener indicates his affinity with von Bertalanffy's views in that he considers social stability to be a form of homeostasis.[66]

He concludes that cybernetic principles are fully applicable in sociology and economics, especially when they acquire the new mathematical methods of physics.[67] The "game" of economics is one in which the rules are continually subject to revisions by the constant flow of technical innovations.[68] The social sciences, says Wiener, are a bad proving ground for the ideas of cybernetics, for the conditions and variables are so difficult to control.

He suggests that the new ideas should first be tested in engineering and biology (sic) before being applied to society. The familiar analogy of the social order to the individual organism is, then, justifiable.[69]

Creative activity, whether in God, man, or machine, is one and the same.

NOTES

1. The following discussion closely follows Ashby's writings.
2. See Chapter IV, on Operations Research, pp. 107–116.
3. W. Ross Ashby, *Introduction to Cybernetics*, p. 40.
4. *Ibid.*, p. 54.
5. *Ibid.*, p. 61.

6. *Ibid.*, p. 71

7. *Ibid.*, p. 87.

8. *Ibid.*, p. 96.

9. *Ibid.*, p. 110.

10. *Ibid.*, p. 110.

11. *Ibid.*, p. 114.

12. *Ibid.*, p. 114–115, 116.

13. *Ibid.*, p. 117.

14. *Ibid.*, p. 127–128.

15. For other definitions of probability, see L. J. Savage, *The Foundations of Statistical Inference* (London: Methuen, 1962), and his *The Foundations of Statistics* (New York: Wiley, 1954).

16. Ashby, *op. cit.,* pp. 165ff.

17. *Ibid.*, p. 168.

18. *Ibid.*, p. 170.

19. *Ibid.*, p. 170.

20. *Ibid.*, pp. 170–171.

21. *Ibid.*, p. 174. The illustration is equivalent to the statistical concept of a sampling space.

22. *Ibid.*, pp. 174–176.

23. *Ibid.*, pp. 189–190.

24. Ashby, like most systems theorists, rarely or never argues that cybernetic systems concepts are applicable to social systems; he merely takes this for granted and simply refers to them in his examples as embodying cybernetics concepts. What we would regard as needing demonstration is never demonstrated but merely assumed.

25. *Ibid.*, p. 198.

26. *Ibid.*, p. 201.

27. *Ibid.*, pp. 203–204.

28. *Ibid.*, p. 211.

29. *Ibid.*, p. 211.

30. *Ibid.*, p. 223.

31. *Ibid.*, pp. 216–218.

32. *Ibid.*, p. 231.

33. *Ibid.*, p. 241.

34. *Ibid.*, p. 244.

35. *Ibid.*, p. 253.

36. *Ibid.*, p. 263.

37. *Ibid.*, p. 272.

38. W. Ross Ashby, *Design for a Brain—The Origin of Adaptive Behavior* (New York and London: Wiley, 1960).

39. *Ibid.*, p. 9.

40. *Ibid.*, p. 18.

41. *Ibid.*, p. 28.

42. *Ibid.*, pp. 102ff.

43. Norbert Wiener, *The Human Use of Human Beings* (New York: Avon Books, 1967;) (originally published 1954), p. 25.

44. *Ibid.*, pp. 20–21.

45. *Ibid.*, p. 36.

46. *Ibid.*, pp. 38–39.

47. *Ibid.*, pp. 46–47.

48. *Ibid.*, pp. 51–52.

49. *Ibid.*, pp. 54–55.

50. Wiener's discussion of semantics cannot be regarded as distinguished or even adequate, either from a logical or an historical point of view. His notion, for example, of the high degree of arbitrariness in human language, though no doubt correct from one point of view, assumes that some codes are less "arbitrary;" they, presumably, approximate to a hypothetical "natural" code, whose symbols are not arbitrary.

51. *Ibid.,* p. 102.

52. *Ibid.,* p. 107.

53. *Ibid.,* pp. 124–125.

54. *Ibid.,* pp. 133–134.

55. *Ibid.,* p. 134.

56. *Ibid.,* p. 150.

57. Wiener does not conclude his chapter on law and communication; he appears simply to abandon the discussion. He does not develop his notion of the way in which cybernetics could be used to render the operation of legal disputes just in the sense that he understands the term.

58. *Ibid.,* p. 243.

59. Pere Dubarle's review, quoted in Wiener, *Ibid.,* pp. 244–247.

60. *Ibid.,* pp. 249–250.

61. *Ibid.,* pp. 263–264.

62. *Ibid.,* p. 264.

63. Norbert Wiener, *God & Golem, Inc.—A Comment on Certain Points where Cybernetics Impinges on Religion,* (Cambridge: M.I.T. Press, 1964), pp. 46–47.

64. *Ibid.,* p. 71.

65. *Ibid.,* pp. 83–84.

66. He evidently assumes that science contributes to what he calls social homeostasis. How it does so is not made clear. Many thinkers have always thought that it was the other way around—that a society that is relatively stabilized, perhaps along traditional lines, is disrupted by the introduction of new technological and scientific devices. Marxism itself assumes that the dynamics of change are explained with reference not merely to the ownership of the means of production, but, below that, to changes in the forms and types of interaction with nature. The words of E. M. Forster express the thoughts of many writers on this subject: "We cannot reach social and political stability for the reason that we continue to make scientific discoveries and to apply them, and thus to destroy the arrangements which were based on more elementary discoveries. If Science would discover rather than apply—if, in other words, men were more interested in knowledge than in power—mankind would be in a far safer position, and the stability statesmen talk about would be a possibility . . . But Science shows no signs of doing this: she gave us the internal combustion engine, and before we had digested and assimilated it with terrible pains into our social system, she harnessed the atom, and destroyed any new order that seemed to be evolving. How can man get into harmony with his surroundings when he is constantly altering them? The future of our race is, in this direction, more unpleasant than we care to admit, and it has sometimes seemed to me that its best chance lies through apathy, uninventiveness, and inertia." From "Art for Art's Sake," in *Two Cheers for Democracy,* (New York, Harcourt, Brace & World, 1951), pp. 90–91. Perhaps Wiener then means stability to be introduced by scientific management of society.

67. *Ibid.,* pp. 87–90.

68. *Ibid.,* pp. 92–93.

Information Theory, Communication Theory, and Artificial Intelligence

The advent of computers, both digital and analog, has seemed to signify for many scientists something more than the appearance of a tool for calculation. The computer is seen as an extension of human intelligence that has the potential of surpassing the intelligence of its creators. Many scientists and popularizers maintain that the computer embodies the laws of logical thinking, without the slowness, ambiguity, emotionality, and capacity for error to be found in the human being.

Accordingly, computers were foreseen that would translate texts from one language to another better and more rapidly than do human interpreters, would play chess on the grandmaster level, and would even display originality and creativity in the discovery and proof of mathematical theorems. Although much of the enthusiasm of the early proponents of computers has evaporated, many writers still make claims for the potential—albeit not yet realized—of computers as artificial intelligences. To this day the computer is seen as a major instrument in decision theory: Models of a social or political structure can be simulated on the computer, the necessary parameters can be specified, and the computer can determine the levels of action that will maximize any specified result. Thus the computer can determine social policy on a scientific basis.

The perspectives adopted by the proponents of the computer as artificial intelligence include the following:

1. *Analysis of Language.* Since language is the vehicle of thought, its properties have been subject to a number of philosophical, mathematical, and statistical approaches, especially with a view to establishing a scientific basis for the problem of meaning in language.

2. *Information Theory.* The transmission of messages over channels such as telephones, telegraph lines, and radio has been analyzed with reference to the mathematical laws governing them.

3. *Communication Theory.* Whereas information theory is somewhat narrower in scope and focuses on patterns of transmission without regard to the meaning of the signals transmitted, communication theory is an attempt to deal with the semantic contents of the signals transmitted.

4. *Boolean Logic.* The laws of logic are considered to be the laws of thought, and the circuitry of the computer is seen as embodying the circuitry of the human brain, though in simpler and cruder form.

5. *Cybernetic Machines Governed by Feedback.* Researchers have constructed mobile machines that seek (or avoid) light, learn to travel around obstacles, and learn to solve mazes and other problems. They are said to learn by experience; some of them play games such as chess, NIM, and checkers. Since their circuitry is considered isomorphic with that of the brain, proponents feel that the construction and study of these machines will provide insights into many unanswered puzzles about the operation of the brain.

6. *Neural networks.* Networks of electrical connections have been constructed that may yield insights into the operation of the human brain and thereby insights into the problem of original and creative thinking.

Some of the workers in these fields have been comparatively modest in their claims, but many have felt the impulse to extend the significance and reach of their work beyond the limits of their disciplines and have written in world-encompassing terms reminiscent of Norbert Wiener, Ashby, and von Bertalanffy. A number of missionary works have appeared in which the systems concept is approached from directions other than that of the open system of biology or the cybernetic concept of feedback. The fusion of views, however, has occurred so rapidly that the distinction between cybernetics and information and communication theory is difficult to maintain. One prominent figure in communication theory, Colin Cherry, says of it that "this wider field is an aspect of scientific method . . . referred to, at least in Britain, as *information theory,* a term which is unfortunately used elsewhere synonymously with communication theory. Again, the French sometimes refer to communication theory as *cybernetics.* . . . It is all very confusing!"[1]

Bearing in mind the overlapping of these various approaches, we try to trace their contributions to the concept of systems thinking.

INFORMATION THEORY

Perhaps the most important document of information theory is Shannon and Weaver's *The Mathematical Theory of Information.*[2] The book consists of two separate essays: a nontechnical exposition by Warren Weaver and Shannon's mathematical theory of message transmission. Shannon is modest in his claims. His work deals purely with the scope and limitations of message transmission, and he explicitly indicates that he is not concerned with the meanings of the messages transmitted, but purely with problems such as the coding and decoding of messages, the preservation of signals in the presence of "static," the practical limits to the speed at which signals can be sent under given limitations, and so on. Weaver, however, is more ambitious in his claims for the significance of Shannon's work, and since the two essays are published together, one assumes that it represents Shannon's views as well.

Communication is broadly defined to include "all of the procedures by which one mind may affect another. This, of course, involves not only written and oral speech, but also music, the pictorial arts, the theatre, the ballet, and in fact all human behavior."[3] But it may also include the procedures by which one mechanism affects another (e.g., a radar device tracking a plane, and guiding a missile toward it). Communication theory is to be defined on three levels:

1. *Level A.* How accurately can the symbols of communication be transmitted (the technical problem)?

2. *Level B.* How precisely do the transmitted symbols convey the desired meaning (the semantic problem)?

3. *Level C.* How effectively does the received meaning affect conduct in the desired way (the effectiveness problem)?[4]

Like many communications experts, Weaver assumes implicitly that *communication* means (or implies) *command*. Thus effective communication means the degree to which the conveyed meaning leads to desired conduct on the receiver's part.

Although the three levels may seem at first glance to have little to do with one another, communications theorists hope to show that they overlap; although the engineering aspects apply only to level A, levels B and C are dependent on accuracies at level A. Limitations of the theory on level A would "necessarily apply to levels B and C."[5]

Weaver then describes the fundamental ideas of communication theory (which, as we have seen, were incorporated by Ashby into his presentation of the basic ideas of cybernetics). These ideas have been described in the preceding chapter. Of more interest are the claims made by Weaver for the wide applicability of the theorems. First, the theory of signal communication is so general that it covers any type of symbol: letters, spoken words, musical tones, pictures. Therefore it deals with the inner core of the communication problem: the basic laws that are generally applicable to specific forms. Thus it is a basic theory of cryptography and is basic to the problem of translation. The basic ideas are so closely connected to the logical design of computers that "it is no surprise that Shannon has just written a paper on the design of a computer which would be capable of playing a skillful game of chess . . . this paper closes with the remark that either one must say that such a computer 'thinks,' or one must substantially modify the conventional implication of the verb 'to think.'"[6]

Furthermore, the basic diagram used by Shannon

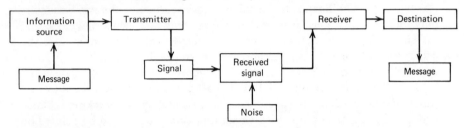

is of such generality as to apply to virtually all situations involving communication, even on levels B and C. The scheme may require minor adjustments or additions. On level B one might want to add a box labeled "Semantic receiver" and another labeled "Semantic noise." Thus a decoder would have to take these into account also.

Third, the theorem that channel capacity cannot be crowded beyond certain limits seems to apply to communications with an audience; if the capacity of the audience is overcrowded, inescapable error and confusion result.

The concept of information developed in this theory at first seems disappointing and bizarre—disappointing because it has nothing to do with meaning, and bizarre because it deals not with a single message but rather with the statistical character of a whole ensemble of messages, bizarre also because in these statistical terms the two words *information* and *uncertainty* find themselves to be partners.[7]

But these reactions are only temporary; Shannon's analysis has so cleared the air that now perhaps for the first time a real theory of meaning can be developed. The next task is to deal with language in statistical terms so that it may be redesigned to say what mankind wishes to say. The study of Markov processes appears to be especially promising for semantic studies, because this theory is adapted to handle one of the most difficult aspects of meaning, the influence of context. "One has the vague feeling that information and meaning may prove to be something like a pair of canonically conjugate variables in quantum theory, they being subject to some joint restriction that condemns a person to sacrifice the one as he insists on having much of the other."[8]

Weaver concludes that the appearance of the concept of entropy in the theory is surely significant; entropy speaks not only the language of arithmetic, but also the language of language. Shannon and Weaver's work represents a landmark for communications theorists, who have since devoted themselves to an attempt to carry its program forward in the analysis of language and thought. Since the appearance of the the *Mathematical Theory of Information* a number of other evangelical documents have appeared by communications theorists; among them are Colin Cherry's *On Human Communication*, Walter R. Fuchs' *Computers—Information Theory and Cybernetics*, Jagjit Singh's *Great Ideas in Information Theory, Language and Cybernetics*, and Donald M. MacKay's *Information, Mechanism and Meaning*.

The main themes of these works are the same—a programmatic description of the basic concepts of cybernetics and information theory, an expression of hopes for new insights into thought and language to be provided by these theorems, and very little or no specific empirical work. The authors remain on the level of programmatic proclamation.

LANGUAGE ANALYSIS

Communications theorists attempt to relate their work to the various schools of modern linguistics. Beginning with the cybernetic emphasis on Markov chains, a vast number of statistical analyses of texts in all modern languages have been attempted. Results appear to have been disappointing.

Markov chains play a part in all situations of communication. This includes "speakin " as a human behavioral characteristic. Is the Markov method also of interest to the linguist and to theoretical linguistics? There is no doubt that in this connection this method has been, and still is, overestimated. Y. Bar-Hillel dissociated himself from this method in a very clear manner: "In the study of language there is no room for probability and statistics." In fact we have pointed out several times that the roots of measurable information do not lie in the landscape of languages; information theory without probability and statistics does not exist.[9]

But attempts to find mathematical laws governing language and meaning continue to be made. One of these is represented by the work of G. K. Zipf.[10]

Zipf held that humans, as goal-seeking organisms, are governed by the principle of minimizing total action between any two points in time; this would apply to mental as well as physical actions and, therefore, to thought and speech. Basing his work on word and letter counts from a vast amount of printed works in many languages, Zipf formulated a linear relationship between the number of meanings a given word has and the frequency of its use. The number of distinct meanings a word possesses is a linear function of its rank order of frequency of use, a relation that seems to apply over different authors and different languages. Words are seen as having "costs" (perhaps in time or in effort); language somehow assigns probabilities to the appearance of words and letters so that their total average cost is minimized.[11]

One of the central concepts of communication theory, drawn directly from Shannon's concept of entropy, has to do with the ways in which communication occurs. The more predictable the message, the less a person communicates; he must depart from predictability in order to say something. Thus the announcement that the sun will probably rise tomorrow, based on so highly probable an event, communicates virtually nothing, while the description of an unforeseen or unusual event is really "news." Communication occurs only insofar as it departs from expectations. But it must not depart too far. The more it departs from predictability, the more original it becomes, but by going too far it becomes incomprehensible. Thus there is, according to Zipf, an equilibrium that must be established between a "force of unification" and a "force of diversification." These concepts, however, have not led to any major insights into the relations between thought and language.

Other quantitative approaches to language have been made in the field of linguistics. The linguistic school of Roman Jakobson has attempted such an approach to spoken language by geometrizing spoken sounds (phonemes). This approach operates by choosing binary oppositions: Sounds may be classified as vocal/nonvocal, consonantal/nonconsonantal, interrupted/continuous, voiced/unvoiced, and so on. These features are represented by orthogonal axes in a hyperspace of 12 dimensions.[12]

But the work of Jakobson, like the important work of Chomsky in theoretical linguistics, seems to have by-passed cybernetics and information theory.

. . . the developments in the field of the exact sciences have shown that the interactions between theoretical linguistics and cybernetics were far less significant than was assumed at first. This must be stated to the regret of many scientists all over the world. . . . The theoretical linguistics of Harris and Chomsky did not require cybernetics, either as midwife or wetnurse, in order to develop as a science.[13]

Nevertheless, the hope is expressed that the two may eventually interact.

One attempt to deal with the meaning or content of language—that of Donald H. MacKay—may be described. Professor MacKay, head of the Research Department of Communication at the University of Keele, is prominent both as research scientist and as popularizer of information theory, and he has contributed to a number of the Symposia on Information Theory edited by Colin Cherry. One such paper indicates his program: "The Place of 'Meaning' in the Theory of Information."[14]

MacKay notes with regret the renunciation of any attempts to deal with meaning made by information theorists such as Shannon. He notes the attempts to measure information content on a probabilistic basis made by Bar-Hillel and Carnap as opening a possible way to the measurement of meaning. This involves the concept of a total "information-space." The measure of information-content is related to the amount of information-space left unfilled by a particular statement. This concept is closely related to Shannon's measure of entropy. But the Carnap and Bar-Hillel approach operates on two levels, a "language of observation" and a "theoretical" (or meta-) language; measures of information content are limited to the latter, while meanings in the "language of observation" are left undefined. (The difference between Shannon's measure of information and the measure of Carnap and Bar-Hillel is based on a different approach to probability; Shannon defines probability in statistical terms, while the latter develop the concept of inductive probabilities; the difference is based on Carnap's approach to language analysis).

MacKay defines *meaning* in formal terms: A message communicated to me may not produce any immediately observable effects on my behavior. What the message is concerned with is my "total state of readiness;" I carry a repertory of possible patterns of behavior for relevant circumstances. I decide which set of circumstances is most likely and which line of behavior is most probably appropriate for a given circumstance (MacKay calls this a "conditional probability matrix"). Thus the purpose of a message is to affect my conditional probability matrix, which in turn governs my behavior.

The fact that a received message may produce no immediate effect on behavior does not mean that it is meaningless. Meaning is not identical to a state it produces, but is identified by the state it produces. Meaning can be defined not as a resulting state of the conditional probability matrix, but as the selective function of the message on the totality of all possible states of the conditional probability matrix. Thus meaninglessness can be defined as the property of a message that lacks a selective function. Statements may lack selective functions for several reasons: Some terms may be undefined, or one term in a statement may be incompatible with another. (MacKay's illustrations of the first are nonsense words; of the second, "The water is isosceles.") This definition helps to explain the meanings of the specific types of statements: questions and commands. A question indicates that my internal switching is not in satisfactory condition. "Its job is to identify the organizing work that needs to be done on the originator's switchboard, so to say. The meaning of the answer will in turn be its selective function—a more detailed one—on the range of the questioner's states of orientation. Its job is to set the switches."[15]

A command or request is defined by MacKay as an indication that action on the part of the receiver of the command is necessary to remove disequilibrium in the originator; thus that "one of the originator's information-flow loops for goal-direction passes through him, so that he is potentially under feedback." The author seems to be aware that his "analysis" is purely definitional and conceptual and offers nothing of an empirical nature. He concludes: "I am painfully aware that the bulk of what needs to be done is little reduced by all this. If it succeeds only in stirring up fruitful discussion in a neglected area it will have served its purpose." [16]

MACHINES AS GENERATORS OF ORIGINAL IDEAS

MacKay's speculations on defining meaning in terms of information theory suggests to him that machines might also act as sources of creative ideas. Digital computers handle logical data and perform logical deductions. They abide by an agreed symbolic code. Can they be conceived of as sources of significant information in the sense of being original or creative? At first glance it would seem not. These machines appear to be mere transducers of information supplied to them by their users; since their output is, at least in principle, completely predictable, they perform no selective actions whatever, and so show no "originality." By contrast, a random noise generator could be regarded as a source of completely original but meaningless information (where the computer is a source of completely unoriginal but strictly meaningful information).

The natural next step is to ask whether one couldn't somehow combine devices of the two kinds to generate processes that are at once original and meaningful. One might create a deterministic logical machine and then connect it to a random noise generator that intrudes upon its functions in selective ways.

If, then, the artefact is equipped to handle meaningful symbols according to the rules of normal syntax, its reactions to a human interlocutor in the field of discussion so symbolized may frequently amount to the generation of new information—new ideas, for example, tentatively advanced, yet normally possessing finite chances of meaningfulness by virtue of the guiding probability configurations under which they have originated.[17]

This compares with Shannon's method of approximating to English sentences by arranging that the relative probabilities of his choices of English words reflect the statistical structure of spoken English to secure a high probability that a random method of choice will produce meaningful English sequences.

But to carry the procedure further, MacKay suggests a device (somewhat similar in structure to Ashby's homeostat) that may learn by experience. The device consists of three or more identical units; the electrical firing of any one unit is random, with feedback indicating to the firing mechanism which unit is more likely to operate first, on the basis of a minimally lower threshold of sensitivity to the buildup of electrical charge. Thus the device will "learn by experience" which unit is more prone to operate first and will tend to favor the firing of that unit. If the responsiveness of the individual units is changed, the artefact will, through feedback, alter the probability of firing any given unit to match the new set of probabilities. The artefact is now playing an optimum game against induced randomness and is therefore in effect symbolizing its (statistical) expectations of what its "target" will be doing next.

Thus one can envisage an artefact "capable of forming 'symbols' for any regularity in the flux of its experience, by adapting its 'state of readiness' to match the regularity." The artefact could grope toward the formation of configurations to match the probabilities of the configurations being offered it.

The artefact will (1) generate actions and (2) generate adjustments of probabilities of actions. It cannot only generate information spontaneously, but can also form symbols for new concepts developed either by incoming data or out of its own spontaneous activity. Thus the likelihood of "meaningless or unprofitable activity would seem in principle to admit of indefinite and automatic reduction."[18] Information is stored on a long-term basis (in the form of transition probability matrices), and the sequences of current internal activity will constitute the symbols of a current "train of thought."

Might human thought processes conform to this model? Might not human originality depend in part upon random physiological processes that are "disciplined" by the "transition probability matrix" that we call experience? McKay suggests that neural processes may in fact be subject to disturbances by factors such as random fluctuations in metabolism. Such events will not in general produce surprising effects, but may on occasion generate "choices not prescribed by the data."[19] These might be new contributions. McKay never makes clear what he means by the notion of "choices prescribed by the data."

To be sure, human discourse is much more complex and much less predictable than this model would suggest. Yet the human being is an information source in two senses: (1) He is a "transducer of information received from outside, from others, or from the abundant information source of the world of physical events." In this sense he is only a secondary source of information. (2) More rarely, output is not completely determined by input, and the result will be both novel and intelligible. Then the human being acts as a "primary information source" in the technical sense of the term.

To be sure, these speculations are rudimentary; we do not fully understand the processing both of original thought and of randomness.

MacKay, like other systems theorists, appears to be laboring under a philosophical confusion that is obscured by his scientific terminology. If by *origi-*

nality or *creativity* we mean the bringing into being of something that did not exist before, it appears to me that he transfers the problem from one level to another without solving it; that is, he appears to assume that original information consists of the combination of preexistent elements in ways that have not occurred before. But if the elements are finite and determinate, so is the number of their combinations and permutations, though the number may be very large. Creativity is defined as the ability to bring about specific permutations of preexistent elements that have not been noticed before. But if the combinations or permutations have not been previously noticed, this does not mean that they have not previously existed, at least in principle. But then originality is defined as the combination of preexistent elements combined with the ability to notice that a specific combination is unusually fruitful or valid. The meaning of originality or creativity has been shifted away from its initial sense of the bringing about of something that did not exist before.

If MacKay (or Ashby and other systems writers who have approached the problem along these lines) does not endow the concepts of originality and creativity with the meaning of the creation of something that did not exist before, but intends by these words the combination of preexistent elements, it appears to me that since the elements are finite and the permutations are also finite, their combinations constitute a closed and specifiable field. Thus the process is an elaborate tautology in that all possible predicates are already contained within the subject. Since by *originality* or *creativity* MacKay probably means the creation of new concepts and formulations, he is further moved to conceive of this process as the combination of words, some combinations of which may prove valuable. This must lead to an analysis of words and their meanings, but I would argue that conceptions of an original or insightful nature do not reside in words or in their combinations and, therefore, do not emerge from them. Rather, the conceptions seem actually to guide the words into appropriate formations. This may often mean the unconventional and metaphorical use of words, which may often strain or stretch their accepted meanings, until the new usage establishes new meanings for the words.

In any event, information theorists seem to avoid examining the processes of thought and invention. Yet creative scientists, artists, and thinkers offer many detailed accounts of their working methods and of the circumstances under which they arrive at major insights. They seem to be engaged in establishing a formal framework within which such investigations are to occur, whether the framework is appropriate to the problem or not. Thus the framework of information theory appears to take precedence over the problem it claims to be able to solve.

ARTIFICIAL INTELLIGENCE

The nature and functioning of the brain and of human intelligence are conceived within equally rigid frameworks, which have been incorporated into the world view of systems thinking. Attempts to understand the brain have been predicated on several levels. The most basic level has been the study of the nerve cell, or neuron, regarded as the fundamental unit of the human nervous system. It has been compared to the transistor in a computer, except that it is much smaller. Artificial brains must use vastly greater amounts of electrical current and occupy greater space than do living brains.

Neural Networks

Attempts have been made by researchers such as Walter Pitts and W. S. McCulloch, Norbert Wiener and Oskar Morgenstern to understand the brain in terms of nerve networks. To the naked eye a neuron looks like a long cord of whitish translucent material. Some neurons are about ½-inch thick; others are so thin they are scarcely visible to the naked eye; some have many branches and offshoots. The joining of the end of a nerve cell to other neurons is called a synapse. Synapses themselves may be very complex. There are estimated to be about 10 billion neurons in the human brain whose inter-meshing forms about 1000 billion internal synapses. This richness of connection appears to permit the synchronous handling of many messages at once and provides security against damage to the nervous system. If some cells are damaged, others may carry the messages instead. Experiments on animals indicate their ability to conduct adequate sensory discrimination of objects with even 90% of the neurons in the brain destroyed.

Computers, though faster than human brains in the performance of simple repetitive computations, cannot carry the richness of sensory messages exhibited by living brains, and they cannot repair themselves. Although it operates on an "all-or-nothing" basis (either "firing" or inactive), and its operation is discontinuous, the neuron can report continuously varying quantities by "counting" the frequency of pulse transmission.[20] But how the brain organizes its incoming sensory input into a coherent experience of the world is not known, and researchers are not near discovery of the process. The topology of the brain is enormously complex. It is a neural network whose 10 billion neurons are each linked, on the average, to 10 others—a total of 10^{12} connections. Its circuitry remains a mystery. Most researchers have been concerned with the gross anatomy of the brain, which sheds little light on its internal functioning. They have discovered that certain regions have well-defined functions. Specific parts of the body can be mapped onto specific parts of the cerebral cortex with respect to sensation and motor action, but these represent barely 1% of the neurons in the cerebral cortex. The deeper interactions within the brain remain an unsolved riddle.

. . . we do not yet know how memory and higher functions are distributed in the overall fine structure of the brain. The brain apparently carries all its immense store of information, some 10^{15} bits, without any storage organ . . . It seems to take all the functions in its stride in an over-all gestalt or intuitive fashion. Apparently all the neurons of the cerebral cortex are in constant activity.[21]

Brain tissue appears to be surprisingly plastic because residual parts take over functions of parts that have been damaged or destroyed. Thus the analogy of brain and computer is problematical; the major researchers in this field, such as McCulloch, Pitts, von Neumann, Turing, MacKay, have explored the possibilities of simplified models of neural networks.

THE CIRCUITRY OF LOGIC

Shannon was the first to work out the physical implementation of logical operations (e.g., addition, multiplication, negation, equivalence) by electrical switching circuits. To each logical operation such as "not," "and," "or," there corresponds a given configuration of switches. The basic device is a switch or contact which is either closed or open; these are called two-state or bistable devices. Other bistable physical devices are magnetic cores, transistors, cryotrons, electron tubes. The bistable *states* or conditions may be conducting versus nonconducting, closed versus open, magnetized positively versus magnetized negatively.[22]

The embodiment of logical operations in switching circuits can be illustrated by the simple operations corresponding to *or* and *and*.

x or y (or both)   x and y

In the first case (x or y), if either x or y is true (switch closed), current flows, and the logical proposition x or y is true. In the second instance both x and y must be true (their switches must be closed) for current to flow, and the proposition x and y is true.

Another electrical equivalent to logical propositions is made possible by an inverted switch—one that is closed unless specifically opened, in contrast to the conventional switch that is open unless specifically closed. Such an inverted switch corresponds to the logical concept: "not-x." Such switches and circuits, properly combined, can be used to embody statements in logic such as "*X* implies *Y*," "either *X* or *Y* but not both," and more elaborate constructs such as "(*A* implies *B*) and (not-*A* and *C*) and (*C* implies) and [*D* implies (*A* and *B*)]."

The analysis of such statements and their corresponding circuitry is conducted by means of "truth tables" which indicate whether the circuit is complete for all possible values (true or false) of *A, B, C,* and so on.

Thus any propositional statement that yields a true proposition (in electrical terms, a closed circuit through which current can flow) for all possible conditions of the variables (e.g., "*P* or not-*P*") is called a *tautology*. Any statement that is not true for all possible conditions of the variables is called a *contradiction;* statements that are true for some and false for other conditions of the variables are called *contingent* statements.

The circuitry developed also accounts for logical functions such as "equals," "greater than," "less than," and the common operations of arithmetic.

Much of applied Boolean algebra is concerned with the search for the simplest circuit patterns corresponding to a given propositional statement. But computer logic developed along Boolean lines is not concerned with the content of statements *A, B,* and so on, but rather with their purely logical consistency. As such, they lead to constructions that everyday human thought could not manage or could manage only with great difficulty; but at the same time they lead to self-enclosed abstract forms that have little to do with human thought in personal, social, or historical contexts. Insofar as they do this, the phrase "machines that think" redefines "thinking" to a highly limited and specialized meaning.

Neural Networks and the Brain

Since neurons in the brain have an all-or none character (firing or not firing), researchers such as Pitts, McCulloch, and von Neumann have suggested that logical propositions could be represented by neural networks as well as by switching circuits. Chains of logical propositions can thus be represented either by computer circuits or by idealized models of the nervous system. The assumption is that concepts are embodied in neural networks and might even be regarded as epiphenomena of neural patterns.

Since the neural mechanisms of stimulus and response are still complex and mysterious, a number of simplifying assumptions had to be made. One assumption is that any stimulus leading to a given response has a unique counterpart in the internal state of an organism. Further, this state must occur *after* the administration of the stimulus and *before* the appearance of the response.[24] Since there must be a time lag involved, there emerges a difference between ordinary logic and the automata designed to embody that logic. While time plays no part in ordinary logic, no real nervous system or model of one can avoid abiding by a definite time schedule. The consideration of temporality requires two further assumptions: (1) that the travel time of an impulse along the axon of a neuron is constant for all neurons, and thus a basic and uniform time unit is assumed for all neural impulses; and (2) that neurons can fire only at discrete moments of time, not continuously. Thus "neurons in our network can fire only when t equals 1, 2, 3 or any other whole number but not when t is 2½ or ¾ or any other fractional number. . . . both assumptions are contrary to fact. They have, nevertheless, to be made to keep the neurons firing in phase and prevent the analysis from becoming intractable."[24]

It is further assumed that neurons may always be classified according to function—that is, there are assumed to be *receptor* neurons that receive impulses from the environment and send impulses to other neurons, *central* or inner neurons that receive impulses from receptor neurons and pass them on to others, and *effector* neurons that receive impulses from neurons and send them to effector organs. Finally, a number of simplifying rules concerning the interconnections of neurons are assumed; for example, one of them states that "each end bulb of a neuron impinges on the soma or body of exactly one neuron, no more, no less."[25]

By means of these constructions, McCulloch and Pitts showed that one could in theory construct nerve networks satisfying any given input-output specifications. The importance of the McCulloch-Pitts work is that it demonstrates that any functioning, of any complexity whatever, can be realized by a neural network, provided it can be defined with strict logic "in a finite number of words." But though they show that such functions are realizable by a neural network, they do not necessarily show us how to do it. In Singh's words:

All that it does tell us is that (*i*) *if* we could get enough neuron cells of various kinds; and (*ii*) *if* they were small enough to be accommodated within the limited space available to the organism; and (*iii*) *if* each cell had enough end bulbs; and (*iv*) *if* we could put enough end bulbs at each synapse; and (*v*) *if* we had enough time to assemble them; and so forth, *then* we could make robots that would behave in *any* completely and unambiguously specified way under prescribed environmental circumstances.[26]

We know of no means to acquire these cells or to make them small enough to be manageable. Furthermore, there is no time to assemble them; an assembly of 10 billion cells, put together at the rate of one cell per second, would require five centuries. Nevertheless, the McCulloch-Pitts theorem is of value in telling us what is possible in principle, though it cannot indicate how to carry this out in practice. Neuron networks, like computer circuits, are based on binary principles (on-off, firing-not-firing), and researchers have shown that any text, no matter how elaborate, can be written in binary code.

As the number of neurons increases, the complexity of their interconnections increases astronomically; thus, with even six neurons the number of networks rises to the 100 billion level. Simplifying devices are used to make these complexities manageable, but they seem to imply a retreat from the real object in its full complexity to a system whose "fit" is almost by definition inadequate.

In any event, the abstract nerve network model can be shown to conform to the basic concepts of Boolean logic in a way completely parallel to the circuitry of the computer. But even the most elaborate networks constructed by theoreticians accomplish but a fragment of the functions accomplished by the brains of even the lowest animals. Nerve-network theory, concludes Singh, is near a dead end, if it has not already reached it.

One attempt to extend nerve-network theory to the point where it begins to approximate the live brain is by von Neumann, who attempted work analogous to Claude Shannon's work on message-transmitting channels. Shannon showed that information could be transmitted across a noisy channel with an arbitrarily high level of reliability by the provision of redundancy. Von Neumann assumed that any neural network has a certain probability of malfunctioning, and he showed that a network of unreliable components could transmit messages with any desired degree of accuracy also by the provision of redundant elements. Von Neumann's mathematical demonstration proceeds by assuming impulse transmission over strands of fibers, not over single nerve fibers. Thus stimulation of fewer than a critical number of fibers does not register as a stimulus; stimulation of more than the critical number will register, even though some fibers do not function with total reliability. But even this simplification does not reduce the number of networks to manageable size, and von Neumann himself warns against identifying computers and nerve networks with the animal nervous system.

TURING MACHINES

The English logician A. M. Turing suggested the possibility of developing machines that could imitate a human being by answering questions well enough to deceive a person into thinking he was conversing with another human being.

The basic importance of Turing's work is that it provided a basis for a machine to be coded to imitate or adopt the behavior of other machines. A

Turing machine consists of a black box, whose internal composition is unknown or irrelevant, but which is capable of a finite number of states; the machine also contains a finite table of instructions for passing from any one of its states to any other. The input into the machine consists of a paper tape of unlimited length, divided into equal squares. In each square may be written either a 0 or a 1. The machine may move the tape either one square to the right or to the left and may either write or erase a 0 or a 1 or a punctuation mark—say, an X. Turing proved that, given a finite set of instructions for writing or erasing and for moving from one state to another (with each state there being associated with a specific act), such a machine could perform any desired computations and any logical operation that could be specified in a finite number of instructions, given a sufficient length of time and tape. More important, Turing showed that such a limited automaton could reproduce the specifications of any other automaton of greater complexity than itself. Since a more complex automaton may be described in a finite number of words, its description could be encompassed by the lesser machine. This description would contain certain empty sections that, when filled in, describe a specific automaton. In von Neumann's words:

As long as they are left empty, this schema represents the general definition of the general automaton. Now it becomes possible to describe an automaton which has the ability to interpret such a definition. In other words, which, when fed the functions that in the sense described above define a specific automaton, will thereupon function like the object described. The ability to do this is no more mysterious than the ability to read a dictionary and a grammar and to follow their instructions about the uses and principles of combinations of words. This automaton, which is constructed to read a description and to imitate the object described, is then the universal automaton in the sense described by Turing. To make it duplicate any operation that any other automaton can perform, it suffices to furnish it with a description of the automaton in question, and in addition, with the instructions which that device would have required for the operation under consideration.[27]

The automaton may operate with as few as two internal states, provided its repertory of tape symbols is suitably increased; or it may operate with only two tape symbols, provided the number of internal states is increased. The principle of the Turing machine has been extended in two important ways:

1. *Computer languages.* Instruction systems have been developed for computing machines that in effect instruct them to act as if they were other machines. Thus, computer languages such as FORTRAN and ALGOL are embodiments of Turing's work.[28]

2. *Self-Reproducing Automata.* Von Neumann's work has indicated that Turing's theorems can be extended—at least in principle—to construct auto-

mata that can produce other automata. Since an automaton can be described in a finite number of words, each part can be specified in its functions. The list can be lengthened by including more elementary parts or shortened by including fewer parts, each of which has more than one attribute or function. The automaton that is to do the constructing is regarded as being placed in a reservoir in which large numbers of the elementary components are floating, and it carries out its construction in that reservoir. Von Neumann suggests that this is in fact how genes operate: A set of instructions for the construction of an automaton and for setting it loose is a gene. A natural gene probably does not contain a full set of instructions, only "general pointers" or clues. A minor variation in the instructions will also be self-producing and will serve as a mutation. Below certain levels complications of organization are degenerative, but above these levels automata can produce not merely other ones less complicated than themselves, but others that are on higher levels of complexity.

COMPUTERS AND THINKING

Though the mysteries of life and thought in their physical bases are being explored further and further, reference to a computer as "thinking" may not represent the same use of the word as when we speak of humans thinking, any more than reference to birds and airplanes as "flying" means that these are really the same activities.

Abstract thinking remains beyond the reach of the computer, which operates on the level of such physical entities as punched cards and electrical impulses. Computers manipulate not concepts but their physical correlates.

Some investigators have tried to create the capacity for abstract thinking in the computer. McCulloch and Pitts developed reading machines that could "recognize" geometric patterns (triangles, squares, circles), but here too the machines are cumbersome and slow. They are programmed to recognize universals such as a circle by computing the ratio of circumference to diameter, and finding this to be a constant through all changes of position, to conclude that it is a circle. The living brain recognizes the same face even when subject to all manner of image transformations by a logic that Singh indicates is different from and beyond the reach of formal mathematics.

Other attempts to liken computers to living beings are based on the recognition that the distinguishing feature of the living organism is self-preservation, and so computer-simulated "personalities" have been developed that classify situations encountered as "frustrating," "painful," or "satisfying."

Further refinements along these lines include the development of nerve-network models with adjustable synapses, which can raise or lower the threshold values of the neurons. This feature is hoped to increase the flexibility of the machines, which will then perhaps "learn" by experience. If a machine receives a signal to the effect that a given response is "favorable" or "unfavorable," it will lower or raise its synapse thresholds accordingly, and in time will "learn" to favor responses that score as favorable. The raising or lowering of threshold response is accomplished by coupling receptor units and establishing a minimum number of such units that must be active for an incoming signal to be recorded; in addition, the units that report correctly are given greater weights than those not so reporting. Thus machines are developed that, it is hoped, can "learn" to recognize visual patterns and to distinguish correctly between different (but similar) letters.

Here too, though a number of procedures have been developed for particular cases, the basic mathematical problems have not been solved.[29] Furthermore, even these methods do not appear to be those actually used by living brains. The number of neurons needed to operate on these lines for the operation of visual patterns alone are more than the 10 billion existing in the human brain. Laboratory rats whose visual cortexes are destroyed have proven still able to recognize food and to avoid obstacles, though other visual performances, such as pattern recognition, are impaired.[30] Furthermore, as Singh points out, feedback-activated machines have proven excessively prone to "runaway hunting," a problem understressed by Ashby.

Thus, Singh indicates, the nerve network approach, while producing interesting but crude automata, still fails to clarify the functioning of natural intelligence exhibited by the living brain.

MACHINES THAT PLAY GAMES

The computer as a problem-solving device proceeds by *algorithmic* methods—that is, it follows a repetitive computational routine until the solution is reached. But in "creative" intelligence such as the generation of valuable work in poetry or philosophy, these methods do not apply. Accordingly, researchers have attempted to solve such problems in mechanical ways by other strategies, sometimes called heuristic strategies. The major area for this sort of work involves machines that play games, especially chess, checkers, and NIM. This approach is seen as nontrivial, especially since games theory is claimed to provide new approaches to economics, sociology, politics, and war.

The rules of each game can be specified by a program, as is a strategy of searching for optimum moves; in addition, some form of "evaluation-function" is provided so that the machines will "learn" from past experiences. Although chess has proven too complex, machines have been programmed to play checkers, a game with simpler rules, at the championship level.

These approaches are considered to be relevant to the problem of understanding human intelligence. The model adopted is that of following the procedure by which a human being would set about proving a theorem in logic or playing chess. In logic one starts with initial premises and the rules of inference; one states theorems and searches for their proof. Similarly, in chess one starts with an initial position of the pieces and rules for moves; one conceives of an advantageous or winning position and searches for the sequence of allowable moves that reaches the desired position.

Programming theorists have sought the solutions to these problems under the heading of "GPS programming" (General Problem Solving). Their components include "goal" components, a learning program, and a performance program. Since the number of possibilities is enormous, the programs operate basically on a trial-and-error basis, and consequently, "GPS programming is fundamentally a recursive process. It selects a subgoal in the hope that it might be easier to achieve than the original goal. It then tries one method after another until the subgoal . . . is achieved. When this happens, it represents a step of progress towards the original goal. A succession of such steps is expected to lead to the main goal." [31]

Some measure of the distance to the goal is provided, with a view to diminishing that distance by each step of the procedure. The procedures involve including a limited number of logical functions; instructions to add or delete variables; increase or decrease numbers; change logical connectives; change signs, groupings, and positions.

But again, the procedures developed are based on the selection and control of finite, logically closed conceptual systems. Thus, while machines have been programmed to prove theorems in symbolic logic by formalistic methods, they show a variety of limitations: Beyond a certain level of complexity they fail to perform; furthermore, they appear unable to discern when to abandon a particular approach; finally, they appear to collapse under the large number of combinatorial possibilities of trials that emerge. So far their achievements have been limited to logic, trigonometry, and algebra. No way has yet been found to instruct machines to abandon the vast number of possibilities in order to proceed directly to a workable result. Finally, although under certain limited conditions machines may be able to prove theorems, none has emerged that appears able to conceive of new theorems to be proven.

UTTLEY MACHINES

A. M. Uttley suggested two possible mathematical principles underlying the organization of the central nervous systems: (1) classification, and (2) conditional probability. [32]

Animals' responses can be evoked by a variety of external stimuli, if the stimuli are similar enough to ones imprinted upon them: Birds may mistake stones for eggs if the shape and color are close enough.

More recent experiments with birds, fish and other animals have conclusively shown that certain combinations of simple properties of the objects suffice in many cases to evoke the requisite response while other aspects tend to be ignored as irrelevant. . . . This is because even though the stimuli emanating from the objects in the external world are in a state of perpetual flux, the mind images they create in the animal's head have in many cases a certain overlap of common attributes that makes them look alike. The existence of this overlap permits the mathematical theory of sets and conditional probability to be applied to design a class of machines which we may call Uttley machines after their designer. They have a structure in some respects similar to that of the nervous system and react to external stimuli in the same way as animals even though to a rather limited extent.[33]

An Uttley machine is constructed of three components: (1) sensor units bringing in signals from the outside world; (2) a central black box whose function is to store signal patterns and to record the relative frequencies of their occurrences, so each incoming signal pattern will be given a measure of relative probability and, further, a measure of the probability of occurrence given the earlier occurrence of other patterns, and (3) output devices.

The nature of the device is such that a pattern is registered only if the number of sensory units operating is over the threshold amount. If the pattern registers below this amount, a quantity is recorded that registers its decreasing probability, thus measuring its rarity.

When the recording unit indicates the occurrence of a given pattern, this rarity function is decreased in value. The rarity function is stored in memory, and all units are linked to one another in all possible combinations.

Uttley's concern is with the possibility that this logical system is embodied by nerve tissue. But the results so far are reported to be inconclusive and not encouraging. Nerve tissue still functions much more rapidly than even the fastest computers in pattern recognition, but even more of a problem is that in electronic devices three separate systems are required for the three functions of input, storage and recognition, and output, while in living tissue all such functions are carried out by the same networks. If animal neurons functioned along the lines suggested by Uttley, many more neurons would have to exist in the living brain than are actually there. To meet this problem Uttley has proposed certain hypotheses to the effect that a conditional probability computer could meet the performance of the living brain through a system of chance connections. As information is accumulated, those connections carrying less information would be eventually disconnected, eliminating ambiguity, while the system improves its ability to discriminate. But such hypotheses remain speculative; Uttley's conditional probability approach embodies the theoretical approaches to statistical significance that were predominant until the 1950s. There is no more reason to assume that nerve tissues operate along these lines than along other lines of approach that could be developed.

THE LIMITATIONS OF MATHEMATICAL THEORIES OF THE BRAIN

As Singh, von Neumann, and others have indicated, there is no theory, mathematical or other, than can correlate specific neuron circuits with particular thoughts, feelings, or states of mind. The reasons are clear: A small "slice" of neuron tissues gives only a microscopic segment of brain processes; a larger section quickly overwhelms the investigator with a vast network of relations that are intractable to analysis. But even if it were possible to know the details of the 1000 billion synaptic connections in the human brain, it would prove irrelevant, because the evidence today suggests that the neural networks of the brain are not "specific" in their functioning. Specific functions are not seated in specific neurons, and if some are damaged, others assume their functions. Furthermore, even the details of behavior that appears to be "built in" to living creatures, such as mating dances and bird songs, cannot be mapped out specifically with respect to genes. And what is involved in the nonspecific behavior of animals is beyond the reach of analytic models. Thus though the fact that a human child can learn a language is no doubt genetically determined, what specific language it will learn is not genetically determined. For such problems the detailed circuitry of individual neurons is probably irrelevant; more relevant, perhaps, are certain general statistical parameters that determine overall patterns of connectivity, but even this view is speculative.

Such theories as have been attempted are at least in reasonable accord with the evidence, but come no closer to solving the problems. Cortical tissue has proven to be too plastic and equipotential to be approximated by the inflexible circuitry of computers.

One recent attempt to escape these limitations was based on the concept of the cathode-ray oscilloscope. The discovery of the spontaneous electrical activity of the brain in the form of brain-wave patterns, variously labeled alpha, beta, delta, and so on by electroencephalographers, has suggested the theory that the brain operates as a scanning device and that it could be understood by recording and analyzing the electroencephalogram (EEG) patterns.[34]

But here too the problems prove intractable to analysis. The complex wave patterns cannot be resolved to any unique set of components. The alpha rhythm seems to be the pattern of the brain when the mind and the senses are at rest; when an object is being perceived or when something is being thought about, the alpha rhythm is replaced by wave patterns that cannot be analyzed. "In sum, . . . no mathematical, statistical or network theory can yet be devised to account for the cortical activity of . . . higher mammals and man. All the various engineering, electrical, and heuristic models devised to mimic or "mechanize" thought processes seem barely to touch the fringe of the problem." The logic and mathematics of the brain seems entirely different from the logic and mathematics of our conscious daily usage. Discovery of its language "still awaits its Newton and Gauss."[35]

Those who wish to construct artificial intelligence on computers appear to be proceeding from bases that are less scientific than philosophical: the perennial mind-body problem of philosophy is "solved" by legislating mind out of existence and by proceeding on mechanistic assumptions. Such an approach is no doubt necessary for detailed scientific work to be done, if in fact the work being done in this field can be called scientific rather than a form of scientific ideology embodied in computer form. The researchers seem to assume that mind does not exist, and if it does, should perform as artificial devices do, and if it does not, must be replaced by these artificial devices. In scientific terms this research has been described as being at a dead end and as having virtually no payoff for biologists interested in the brain.

Nevertheless, although computers are still tied to algorithmic processes, although "perceptrons" show rather poor ability to discriminate patterns, and although living brains appear to operate by a logic not even approached by mathematical logic, which appears to be unfathomable to brilliant mathematicians such as von Neumann, the claims made by men in this field are popularized among wider audiences to the effect that the secrets of artificial intelligence have been discovered.

At this point their work becomes of interest to the sociologist. We discuss the sociological import of these claims in Part III of this study. Here let us say that the claims made by these researchers remain largely programmatic and unfulfilled, in spite of much genuine and brilliant work in narrow and well-defined areas. But all claims to have penetrated the secrets of the brain, much less of the mind, or to have developed intelligent machines in any but the most artificial meaning of the word *intelligent*, must be discounted.

NOTES

1. Colin Cherry, *On Human Communication—a review, a survey and a criticism*, 2nd ed. (Cambridge: M.I.T. Press, 1966), p. 217.
2. Claude E. Shannon and Warren Weaver, *The Mathematical Theory of Information* (Urbana: University of Illinois Press, 1949; reprinted 1972).
3. *Ibid.*, p. 3.
4. *Ibid.*, p. 4.
5. *Ibid.*, p. 6.
6. *Ibid.*, pp. 25–26.
7. *Ibid.*, p. 27.
8. *Ibid.*, p. 28.
9. Walter R. Fuchs, *Computers—Information Theory and Cybernetics* (London: Rupert Hart-Davis, 1971), p. 173. The American edition of this book is identical. *Cybernetics for the Modern Mind* (New York: Macmillan, 1971).
10. G. K. Zipf, *Human Behavior and the Principle of Least Effort*, (Cambridge, Mass.: Addison-Wesley, 1949).
11. Colin Cherry, *On Human Communication*, 2nd ed. (Cambridge: M.I.T. Press, 1966), pp. 102–109.
12. *Ibid.*, pp. 95ff.
13. Fuchs, *op. cit.*, p. 48.
14. Originally published in *Information Theory* (London: Butterworths, 1956), and reprinted in Donald M. MacKay, *Information, Mechanism, and Meaning*, (Cambridge: M.I.T. Press, 1969).
15. MacKay, "The Informational Analysis of Questions and Commands," *op. cit.*, p. 100.
16. *Ibid.*, p. 104.
17. MacKay, "Generators of Information," *op. cit.*, pp. 132–145.
18. *Ibid.*, p. 142.
19. *Ibid.*, p. 143.
20. Jagjit Singh, *Great Ideas in Information Theory, Language, and Cybernetics* (New York: Dover, 1966), p. 137. Much of the discussion in this chapter draws heavily from Dr. Singh's summaries of work in these fields.
21. *Ibid.*, pp. 143–4.
22. Franz E. Hohn, *Applied Boolean Algebra*, 2nd ed. (New York: Macmillan, 1968), p. 70. The important documents include: C. E. Shannon, "A Symbolic Analysis of Relay and Switching Circuits," *Transactions of the American Institute of Electrical Engineering*, Vol. 57 (1938), pp. 713–23; C. E. Shannon, "The Synthesis of Two-Terminal Switching Circuits," *Bell System Technical Journal*, Vol. 28 (1949), pp. 59–98; J. Riordan and C. E. Shannon, "The Numbers of Two-Terminal Series-Parallel Networks," *Journal of Mathematical Physics*, Vol. 21 (1942), pp. 53–93.
23. Singh, p. 147.
24. *Ibid.*, p. 148.
25. *Ibid.*, p. 150.
26. *Ibid.*, pp. 152ff.
27. John von Neumann, "The General and Logical Theory of Automata," reprinted in Walter Buckley, ed., *Modern Systems Research for the Behavioral Scientist*, (Chicago: Aldine, 1968), p. 105.
28. Singh, *op. cit.*, p. 195.
29. Singh, p. 239.
30. *Ibid.*, p. 246.
31. *Ibid.*, p. 270.

32. A. M. Uttley, "The Probability of Neural Connections," 1954; "The Conditional Probability of Signals in the Nervous System," 1955; see Colin Cherry, *op. cit.*, pp. 324, 327.

33. Singh, *op. cit.*, p. 296.

34. *Ibid.*, pp. 314ff. See also W. Grey Walter, *The Living Brain* (Baltimore: Penguin, 1961).

35. Singh, *op. cit.*, pp. 323, 324.

Operations Research and Systems Analysis

O perations research and its offspring, systems analysis, are the most important of the systems approaches, having had a direct impact on American government and policymaking at the federal, state, and municipal levels. They have probably received more publicity than other systems orientations, such as those of von Bertalanffy and the cyberneticians, which have contributed more to the philosophical pretensions of systems thinking than to its effects on public policy.

Operations research is the older of the two and consists of a large and diverse set of "cookbook" approaches to narrow and highly specific problems; systems analysis represents the same approach when it moves beyond the confines of the assembly line, the automated factory, and the weapons system to seek the solution of larger societal problems such as urban transportation, pollution, the delivery of health services, and the planning of wars on poverty. Operations research focuses on the limited and the specific; systems analysis is more global in its aims, though here as elsewhere in considering the entire array of systems approaches the drawing of hard-and-fast lines of demarcation must be somewhat arbitrary and tentative.

OPERATIONS RESEARCH

THE EMERGENCE OF OPERATIONS RESEARCH

Operations research had its beginnings in Great Britain during the late 1930s in the preparation for World War II. The invention of radar led to the development of efforts to use radar devices in a coordinated way and to cope with enemy efforts at jamming the radar. Such an approach was in effect the first "systems" approach: Rather than employ installations as individual units, they were to be deployed and integrated from a "total" point of view.

The men employed in this research were a combination of military officers, civilian scientists, and government officials. As time went on, operations research, in the form of the application of statistical methods to military problems, spread from work on radar systems to the analysis of fighter losses in France, the analysis of aerial bombing raids, the evaluations of weapons and equipment, and to the analysis of specific tactical operations. Foreshadowing its expansion still further, the methods of operations research were addressed to predicting the outcomes of future military operations with a view to influencing policy and to the study of the efficiency of the organizations that deployed equipment and weapons in battle.[1]

The operations research approach rapidly spread among British military and naval commands and was soon adopted among United States commands. During and after the war both the idea and the organization of operations research spread rapidly both to universities and to private industry. Especially from the Massachusetts Institute of Technology the gospel spread to large firms in the oil, electronics, and aircraft industries.

In 1957 the International Federation of Operations Research Societies (IFORS) was founded at a conference at Oxford. This date may be taken as symbolizing the emergence of this approach as a significant phenomenon. Since that time the number of persons employed in operations research as well as the number of organizations and publications have multiplied rapidly.

From the beginning operations researchers (or "systems men") have been active missionaries, convinced that their technique could be applied to a wide range of social and political problems.

As of the beginning of 1968, there were twenty-one member societies of IFORS, representing a membership of over 12,000. . . . I know of 14 journals devoted primarily to the field, . . . and I have probably have not counted them all. The *ORSA Abstract Journal* shows that papers on operations research are appearing in increasing numbers in other journals as well. The conversion is proceeding apace.

But we have some missionary work still ahead of us.

Our next big job of infiltration is into the field of urban and regional management.[2]

Systems analysts have set their sights both higher and wider, and systems approaches made serious claims in fields such as transportation, waste disposal, health, education, social welfare, foreign policy, crime control, traffic control, psychotherapy, population control, food supply, futurology, as well as the understanding of life and of the universe.

Systems men have produced, in addition to purely technical writing, a large literature of missionary works for the reading public, explaining something of the technical apparatus of operations research, and making claims for their universal application. Despite its size, this literature is repetitious and exhibits only a handful of ideas and assumptions. These ideas and assumptions include the underlying concepts of classical statistics and of positivist philosophy (though the latter is often not identified). What little variety there is in the literature emerges in the illustrations used rather than in the thought, which exhibits almost complete conformity to the notion that abstract conceptualization constitutes both a guide to and a substitute for concrete experience.

THE TECHNIQUES OF OPERATIONS RESEARCH

"Operations Research is the application of scientific methods, techniques and tools to the problems involving the operation of a system so as to provide those in control of the system with optimum solutions to the problems."[3] ". . . I would define it as the study of administrative systems pursued in the same scientific manner in which systems in physics, chemistry and biology are studied in the natural sciences."[4]

Operations research is a science of administration. Its techniques include statistics, simulation and Monte Carlo methods, linear programming, queuing theory, decision theory, games theory, simulated games, cybernetics, and information theory.

CLASSICAL STATISTICS

The full apparatus of statistical sampling and testing is applied to factory operations of all sorts including quality control; the testing of the greater efficacy of one machine, drug procedure, and so on as compared to another; the sequential sampling of manufactured product; work sampling applied to time and motion studies. Statisticians working on the most substantive levels such as quality control are not yet systems analysts; when they apply these and related techniques to administrative processes, however, they in effect become systems analysts.

LINEAR PROGRAMMING

Linear programming consists of a set of techniques for maximizing output and minimizing costs under complex conditions. The term *linear* refers to the use of linear equations (the productivity of a given machine is stated in linear equations, x units per hour, $2x$ units in two hours, etc.). Thus a machine shop in which there are eight machines of various states of age, repair, and productivity, and to which 20 or 30 production jobs come per month, is faced with a vast number of possible ways to schedule the routing of production jobs and their subphases among the machines. Most shops operate by traditional or improvised methods; linear programming attempts to plan how to route the production jobs and their components among these machines in the most expeditious ways.

The linear programmer will develop charts of the production equation of each machine and will seek, by geometric superposition of these equations, to find the point of optimum profit—that is, the combination of machine use that is maximally productive.

Linear programming concerns itself with optimal solutions for situations where the number of such units is very large. It is applied to a wide range of problems: What combinations of oil tankers of differing speed and capacity are to be dispatched to a refinery at what specific times, given the productive and storage capacities of the refinery, so as to maximize the flow of oil and to minimize tanker idleness and to avoid the need to shut down a refinery due to lack of storage capacity? What goods are to be dispatched to a given customer from which warehouse, with a view to minimizing delivery mileage, customer waiting time? The technical problems of linear programming arise from its setting up of large numbers of linear equations having hundreds of variables and hundreds of limiting conditions, and it becomes dependent upon the existence of computers with large capacities and elaborate programming techniques.[5]

QUEUEING THEORY

Queueing theory deals with techniques for reducing waiting times under a variety of conditions: more telephone calls may flood a switchboard or an exchange than can be handled at once, and a backlog develops. So many airplanes may arrive at an airport that some must circle about in holding patterns until cleared for landing. Similarly waiting times may develop for customers at a ticket office, drivers at a bridge or tunnel, and so on. Formally similar are problems involving stocks kept in inventory for which sudden demands may be made. The approach to such problems is statistical in nature. Averages are computed (average waiting time, average time between arrivals, etc.), and experimental approaches are adopted whose purpose is the statistical minimization of waiting times. Solutions involve the provision of reserve capacity within the limits of allowable expense.

SIMULATION, MONTE CARLO METHODS, AND OPERATIONAL GAMING

The computer simulation of complex industrial processes is perhaps the most widely used and most highly praised technique of operations research, and one for which far-reaching claims are made. Where situations are too complicated for mathematical analysis and where data cannot be expressed in mathematical form, simulation techniques have emerged.

Computer simulation is based on the idea of operating a model or "simulator" which is taken as a representation of the system. The model is subject to manipulations that would be too complex or expensive to perform on the real system. The operation of the model is studied with a view to its reference to the system under study.

Simulation is in some respects an outgrowth of decision theory, especially where the number of decisions to be made grows too large. The stock device of the decision theorist for demonstrating the proliferation of decisions is the "decision tree."

Thus a firm faced with the option to make or not to make a proposed product in bidding for a contract must trace a network of outcomes and decisions:

1. To make/not make the product.
2. Product meets/does not meet specifications.
3. If product meets specifications, is/is not selected for contract.
4. If product does not meet specifications, redesign/abandon project.
5. If decision is to redesign: (*a*) product meets specifications and is/is not selected for contract; (*b*) product does not meet specifications.

The "decision tree" corresponding to the above might be this:

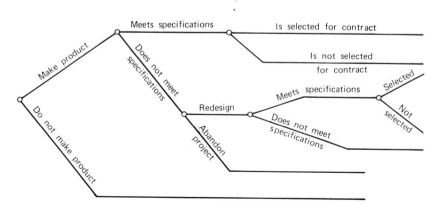

For each branching point of the decision tree, quantitative values are assigned in the form of:

1. An assigned probability of outcome.
2. An estimate of corresponding costs and profits.

Accordingly, outcomes can be assigned expected values, positive or negative.

But in many situations the alternative possibilities are too great in number,
and actual costs can be estimated only within a certain range, due to variable
factors such as changing costs of materials or of labor and uncertainties as to
the real length of the production process.

At this point the decision tree is abandoned in favor of the simulation
process.

Sampling of Factors for Simulation

In creating a model for simulation purposes, many of the relevant variables
may assume values that cannot be known in advance and must be estimated.
Some essential variables, such as an estimate of demand for a product, may
range within broad limits. Thus demand for a product might be estimated as
running between 500,000 and 700,000 units. For simulation purposes, this
range will be broken up into 10 class intervals, and an estimated probability
value will be assigned to each class interval (such that the total probability =
1). Similar appropriate ranges will be assigned to all other relevant factors to
be entered in the simulation model, such as levels of investment, production
costs, amounts to be produced. The unknown and presumably random variations among all these variables will each contribute to the final outcome.
Where there are a large number of factors, each of which may vary over a
range, the simulation process involves sampling each of these variables on a
random basis. Since random processes are embodied in such devices as tossing
coins, throwing dice, or spinning a roulette wheel, they are referred to as
Monte Carlo methods.

Thus when one value for each factor has been randomly selected, an outcome is generated by the appropriate equation that relates the factors of total
demand, level of investment, the firm's share of the total demand, and the
probable price and cost factors.

Simulation programs are used to select randomly an estimated value for
each factor and to develop the outcome. A large number of such "simulation
runs" is generated, each varying randomly, and so a total profile of most
probable outcomes is generated. On the basis of such outcomes, probability
estimates are made of the likely range of return on investment.

The principal concern of the expert in simulation is the internal logic of
the simulation process as it is embodied in a "flow chart" and a suitable

computer program. Much of the craftsmanship of simulation involves constantly refining the simulation program to produce closer correspondence to the complexities of the process being simulated.

The simulation model itself consists of a list of variables, of the relations between them and usually a statement of constraints upon key variables beyond which they cannot range (e.g., avoiding negative amounts of inventory).

Much of the work done in the theory of simulation involves the search for mathematical models—such as the normal curve, the Poisson distribution—to which real processes are held to conform. Computer simulators then build these models into computer subroutines as representatives or images of these empirical processes.

The advantages of the simulation model involve the ability to manipulate the model (change number and types of machines, number of men employed, work schedules) and see what "predictions" the model makes with respect to costs, speed of completed work, bottlenecks, idle time, and so on. Such simulated results can be produced by computers in great numbers, while in practice only a few could be attempted at any time, and if wrong choices were made, could be disastrous; computer outputs would indicate the "wrong" results without actually incurring the penalties.

Operational Gaming

An activity closely related to computer simulation, operational gaming[7] is similar in structure to such board games as Backgammon and Monopoly. An operational game is played by individuals or teams in competition. The board may represent a city, state, nation, or portion thereof designated as a sales region. The players are advised of constraints on the game, such as the number of salesmen they may hire, budget allowed for research and development, advertising amounts. They are free to make decisions within these limits about the number of salespersons to hire, where to send them, pricing policies, amounts to spend on advertising, plant capacity, and the like. They make decisions and place markers on the board representing these factors.

Tables of random numbers are used to establish external conditions such as personnel turnover, probability of a sale per 100 sales calls, and actual market demand. The results of team decisions, which must be made within specified time limits, are computed either by desk calculators or through computer programs, and the players learn the results of their decisions in terms of sales and profit-and-loss statements. Thus operational gaming represents a simulation model in which individual or team decisions are made and serve as inputs. Its principal use is as a training device for managerial personnel.

Operational gaming in business is an outgrowth of tactical war games, which are also simulated on boards and tables representing maps, terrains, nations, such as the RAND Corporation's STRAW (Strategic War Games).[8]

NETWORK ANALYSIS FOR PLANNING AND SCHEDULING (PERT)

As Duckworth indicates, network analysis is known by many names: PERT (Program Evaluation and Review Technique), CPM (Critical Path Method), PEP (Program Evaluation Procedure), LESS (Least Cost Estimating and Scheduling), SCANS (Scheduling and Control by Automated Network Systems), and many others. PERT and network analysis techniques are probably the central techniques of operations research.[9]

To some extent network analysis can be described as time-and-motion study techniques applied to complex administrative systems for specific tasks of production. The basic approach is analysis of a task in terms of the time and personnel necessary for each stage. The primary concern may be to plan the allocation of resources among the stages of the work to minimize slack time and duplication of effort. Of central concern is the idea of the "critical path," which is normally a diagram of steps the task must follow. Thus if assembly B requires the existence of assembly A already completed, resources of men and equipment must be allocated to assure that the output of assembly team A is timely with reference to the next stage.

Network analysis takes cost factors into account, and appropriate work schedules are developed according to limiting factors such as (1) time being all-important and cost no consideration, or (2) working to the optimum time schedule within fixed cost limits.

In programs such as developing the Polaris missile and the space program, involving hundreds of organizations working on specific phases of a program, the network analysis diagrams become vastly complex, especially where uncertainty in production times is inescapable. Statistical methods and computer programs have been developed to calculate critical paths under various circumstances of varying probability. "Its use is being extended into the planning of research programs involving uncertainty of outcome of decisions."[10]

OPERATIONS RESEARCH TECHNIQUES IN GENERAL

Operations research is eclectic in its techniques, borrowing techniques everywhere—from statistics, actuarial theory, Boolean algebra, symbolic logic, queueing theory, computer programming, linear algebra, games theory, cybernetics, information theory, and others. In fact systems persons claim as "techniques" approaches that have not yet been used in specific instances of operations research, such as information theory and cybernetics.[11] They are eager to indicate that operations research

. . . is not just a collection of techniques, it is . . . an attitude of mind of the experimentalist who wishes to find logical causal relationships, albeit statistical ones, in the object of his study and then to use the results of his observations to gain greater control of the environment.[12]

Systems analysis . . . is still largely a form of art and not of science. An art can be taught in part, but not by means of definite fixed rules which need only be followed with exactness. Thus, in systems analysis we have to do some things that we think are right but are not verifiable, that we cannot really justify, and that are never checked in the output of the work . . . we must accept as inputs many relatively intangible factors derived from human judgment. . . . Wherever possible, this judgment is supplemented by inductive and numerical reasoning, but it is only judgment nonetheless.

Our hope for guidance is to turn to science. The objective of systems analysis and operations research . . . is primarily to recommend policy rather than merely to understand and predict.[13]

The view of operations research as partly art, partly science, pervades the literature. Nevertheless, in its practical applications operations research is focused and disciplined in its approaches by the specific requirements of an industrial process or a marketing problem. Under such circumstances it is highly specific, narrow, and technical.

When the operations researcher turns his gaze to wider fields and becomes a "systems analyst," he begins to make social, political, economic and bureaucratic claims.

SYSTEMS ANALYSIS

SYSTEMS ANALYSIS AS A "DISCIPLINE"

Systems analysis is described in its textbooks and missionary documents as both as old as the pyramids and as new as the Space Age and as indispensable if we are to acquire knowledge of relevant and dependable relationships in a complex world.[14] For the systems analyst, scientific method *is* systems analysis, and systems analysis is scientific method. In a narrower sense systems analysis refers to the diagnosis, design, and management of complex configurations of men, machines, and organizations. To design, understand, and improve such systems, the systems analyst must have almost universal knowledge, not only of engineering, but of biological, medical, and social systems. The systems analyst claims to master and combine mixed disciplines and to exploit the gaps between the historically evolved disciplines of knowledge. To improve business and social systems, the systems analyst must be free to disregard formal boundaries and to cross them at will. Systems analysis is a method of evaluating and choosing organizational and social goals and means.

In constructing systems analysis as a discipline, the systems theorist draws heavily on the conceptual apparatus of cybernetics, information theory, the open system of the biologists, and computer languages. The core of a system is a diagram or a flow chart that consists of elements and the relations between them. The relationships may be termed transactions, interactions, inputs, transmissions, connections, links, and so on. The elements may be called nodes, components, operations, vertices, and so on. Arrows are used to indicate inputs and outputs. Such diagrams represent the system under study. A system element may be a quantitative or logical relation or a set of variables under a transformation, not merely an elementary unit. Such systems are studied by recording the history of its inputs and corresponding outputs, precisely in the manner described by Ashby.

For systems analysts, virtually any diagrammatic representation, such as a family tree or a pyramid representing power in an organization, is a system description. Such systems representations are claimed to have evolved historically. For the Egyptians, power was represented by pyramids; later when machines became important, an organization was seen, in mechanistic terms, as a group of wheels and interlocking gears; with the emergence of the knowledge of the circulation of the blood, new concepts of flows, channels, storages, and delay have appeared. Today we see the emergence of teleological or purposeful systems; the cybernetic view of the world is now seen as dominant.

THE BLOCK DIAGRAM AS TOOL OF SYSTEMS ANALYSIS

The central formula of systems analysis is the block diagram, with arrows representing inputs and outputs. Common types of such operations include:
1. The transformation block, in which input x is converted to output y by specified conversion rules.

2. The decision block, in which input x is tested and converted either to y or to z according to stated rules.

3. The feedback block, in which input x is modified as a function of output y.

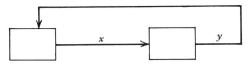

The three steps represent, respectively, conversion operations, logical operations, and correction operations. These are combined and multiplied to any desired level of complexity.

Each transformation point may be conceived as a black box in the Ashby sense. How it is constructed or how it performs the transformation does not matter; its interior is unknown, and it is tested by inputs and the observed corresponding outputs; it presupposes only stability and consistency.

Given these, all logical operations can be developed—loops, decision trees, feedback diagrams, operations with time lag, and so on.

THE FLOW GRAPH

Many systems can be described by means of flow graphs in which the variables become the elements or nodes and the transformations become arrows. Following certain conventions—that all arrows leaving a node carry the variable at the node, and that the value of the variable is the sum of all the transformed variables entering the node as signals from incoming arrows, and that a variable at an arrowhead equals the variable at the tail of the arrow multiplied by the transmittance—the laws of multiplication and addition can be applied to such diagrams. In a similar manner, the black boxes of a system are shown as rows and columns of a table, and the variables connecting them are shown as table entries. The transition of a machine from one state to another can be diagramed in the manner of transition trees described earlier.

Thus for visual presentations, systems analysts adopt mathematical trappings, although these are never applied consistently to any great degree.

SYSTEMS HIERARCHIES

Systems may be ordered into hierarchies of broader and more complex systems or may themselves be broken into simpler subsystems. In systems analysis terms black boxes may contain black boxes or may be contained in larger ones. But what systems analysts appear to mean by system hierarchies also includes the classification schemes used to organize data and files, the Dewey decimal system for the organization of libraries, IBM code systems, as well as the "natural" hierarchies represented by the system of atomic weights. Thus a "system" consists of a set of categories and rules for classification of entities, their placement within specific hierarchies, and the recording of their attributes. The goal of such classification systems is controlling them by means of computerized rules. Corresponding to the varieties of types to be found in the system to be controlled, the computer develops hierarchies of classification and response. At this level the systems analyst discusses the control of a dynamic system, without always indicating what the system may be. It may include a parts inventory for a factory, a payroll system in a computer, or a manufacturing process. The problem, then, becomes one of developing systemic analogies, within computers and computer processes, corresponding to formal classifications used in factory and office processes.

Here systems analysis acknowledges its debt to cybernetics, and adopts from cybernetics the hierarchical scheme used by Wiener and Ashby to describe and classify machines according to the type of feedback present in the machines: (1) First-order feedback includes those machines in which simple adjustments are made—for example, the thermostat that turns the furnace on or off as the temperature fluctuates; (2) second-order feedback includes machines that contain memory storage features for recording past changes of conditions, which enable the machine to react not merely in a "blind" or mechanical way, as with first-order feedbacks, but also by reacting to other past changes of input conditions with preexistent alternative plans. This is equivalent to the step functions described by Ashby.

A great deal of the detail work of systems analysis consists of devising and programming the symbols and codes to be used in computers. At this point, systems analysts adopt wholesale the theory and methods of information theory—if not always in practice, at least as part of their "philosophy" of procedure.

Control systems include (1) *memory media*—card files, magnetic cores, and so on, (2) *memory organization*—types of memory storage that include data storage, pattern storage (in which rules of operations are stored), and (3) *learning storage*—stored instructions that provide for changes in goals and procedures, depending on inputs. Although stored instructions are fully determined, systems analysts, like cyberneticians, argue that such step functions enable machines to learn by experience. The simple machine with feedback is a crude determinate machine, the machine with memory is further along the hierarchy, and the machine with step functions can select (preexistent) plans and develop predictive feedbacks. Thus systems are claimed to learn and change goals even while performing preset lower-order functions.

Like all other disciplines that are indebted to cybernetics, the systems analyst pays homage to Ashby's Law of Requisite Variety: The systems analyst must have at his disposal at least as great a variety of possible actions or resources as the system he wishes to control. For the most part, the systems analyst seeks to control a system by formalistic, statistical methods—that is, he hopes to encompass the essential features of a system in a list of variables whose performance is summarized in statistical measures, such as means, variances, and mathematical distributions such as the normal, Poisson, binomial, multinomial, and others. This approach is not unified, but thoroughly eclectic: Geometrical diagrams are presented as models, algebraic formulae, contour maps, flow charts, normal curves; they are theoretical or systematic, with no unification.

Of similarly loose character is the systems analysts' portrait of values and decisions. Systems analysts claim to be able to evaluate both goals and the measures taken to achieve them. Difficulties emerge because of different and inconsistent criteria of effectiveness; strategies can be evaluated in conflicting ways. In addition, goals of a system may prove to be mutually incompatible or to generate contradictory pressures on those seeking to achieve them. Where ambiguities of this nature arise, the systems analyst attempts to arrive at an "optimum" solution. But of course the criteria for measuring system effectiveness vary. Thus a firm's profitability may be measured by cash liquidity, minimum bank charges, minimum number of bad accounts, and so on; but each of these criteria may generate conflicting actions within the system. The effectiveness measure for a particular department, then, must be harmonized with those for all other departments.

At this point the systems analyst seeks to rise above the system, to see it from a detached point of view, and to resolve the conflicts and contradictions. The implications for his claim of a privileged position vis-à-vis the system are clear.

Suppose an investigator were assigned the task of finding the quickest way to control a new job in a large-scale organization. (This is the task confronting a new administrator unfamiliar with the details of an organization.) Suppose further that, after cursory examination, the organization itself is found to have many of the goal-conflict problems we have described, as well as numerous political factions, operating difficulties, wasteful practices, and technical inadequacies. How should the investigator begin? What questions should he ask? How could he cut through oceans of data without losing perspective? If he is to supervise, how does he obtain control of his job and his organization?[15]

He will start with a higher-level approach.

Although he may be expected to find only a handful of powerful individuals in even the largest organizations, and also to realize that his main coercion sources lie in the power of the budget and the power of appointment, he may choose at first to shun these individuals or devices and view the organization at "arm's length." So that he may not be confused by too close a view, or be swayed by the briefings or political charm of the organization's present department heads, or waste time in a continuous series of conferences, he may prefer to formulate and to ask a series of pertinent questions in writing, and to delegate responsibility for answering them to particular individuals, with particular deadlines.[16]

Based on the answers received and not received, he will spot deficiencies, conflicts, failures, and will demand departmental evaluation reports. He will redefine the structure's goal hierarchy; he will restructure departments, relying on his budgetary and appointive powers. Systems analysts are fond of citing various "triumphs" of the systems approach. Hare, like others, refers to the reorganization of the Department of Defense under Robert S. McNamara beginning in 1961. His simplification of the Department was implemented by creating new functions—the Defense Intelligence Agency and the Office of Organizational Management. Change was achieved not by changing the practices of the military directly, but by changing the measures and questions by which their goals were evaluated. The invasion of the Defense Department by systems approaches has been somewhat less than a triumph. Equally dubious are other "successes" claimed, among them the operations research decision during World War II to focus Allied bombing efforts on the German ball-bearing industry; not many today would claim that aerial bombardment of German industry shortened the war. Hare's reference to the reorganization of the Chicago police force after 1960 by O. W. Wilson, ("like McNamara, a former university professor"), is equally unfortunate.

DATA BANKS

To undertake systems analysis, sometimes also called capability studies, vast resources of information, stored in computerized form, are claimed to be necessary. But when one arrives at the summit of systems analysis as an art, craft, science, or philosophy, one discovers a vague potpourri of questions, formulas, and diagrams, suggesting that systems analysis is a vast cookbook that includes whatever systems analysts happen to do.

The systems analyst calls for the collection and storage of enormous amounts of information necessary for the analysis of systems operations. The main concern is with the development of all-inclusive categories and file systems and with the problems of their coding, storage, and retrieval. As operations are performed, they are coded and recorded. The management information system becomes the controlling brain of the organization.

An integrated management information and control system may be defined as a management-oriented system conceived and designed by management as a single, total entity to control an entire organization. It does not evolve as a result of the development of many more or less independent applications . . . but with the needs of the whole organization in mind. This results in a system welded together by data flows where redundancy in data storage and the transmission of useless information from one area to another are eliminated.[17]

For many systems experts, the major output of factories and business organizations, along with whatever goods and services they provide, is information to be stored on computer tapes and discs for future use. A few systems thinkers, Churchman among them, question whether "facts" exist apart from specific problems or needs; but for the most part the concept of a central data bank is a basic tenet of the systems approach.

SYSTEMS ANALYSIS AS A DIAGNOSTIC TOOL

Systems analysts, then, see themselves as performing a function of coordinating and clarifying goals of a total system from a generalist point of view, which individuals working on specific tasks are likely to overlook, occupied as they are with specific limited goals such as the production outputs of a specific department or subsystem. Systems analysts ask questions: What is the system? How does it operate? Does it work as predicted? Can it be improved? What are the effects of treatment and change? Within this context, simulation studies play a central role. But again, the approaches to these problems prove to be highly unsystematic, involving a grab bag of techniques and tools that rely heavily on quantitative, statistical, and computerized approaches.

When the systems approach turns to broader social questions, it emphasizes a concept of man and society in behavioral terms. Van Court Hare's text, like Norbert Wiener's view of society, may be regarded as a representative document.[18] He adopts as his model of society a scheme taken primarily from Allen Wheelis' *The Quest for Identity*.[19] Wheelis distinguishes between two fundamental social processes: the instrumental process, which represents scientific rationality, respect for "the facts," and the growing mastery of man and nature which results, and the institutional process, which by contrast, is the source of values and of certainties. Religion, rather than science, is the model offered; but the institutional process is seen as wider than this; it includes customs; taboos; and social institutions such as kinship, the state, and private property, which are accorded unconditional validity by men and societies. It is the source of individual and social conscience. Hare, like Wheelis, considers these two categories to be in opposition. The instrumental process brings about change; the institutional process, which apparently represents a residue of earlier instrumental achievements, resists change and accepts it slowly. Thus the phenomenon of cultural lag: the gulf between what is possible and what is acceptable. The problem for the systems expert, then, is introducing change in spite of institutional resistances. The techniques to be used are manipulative where they are not overtly authoritarian.

Change is easier to introduce in matters arranged on a scale with narrow intervals than in those arranged in a sharp dichotomy . . . is easier to introduce through existing institutions rather than through new, through individuals of high prestige and status rather than low, through a third "disinterested" party rather than directly . . . the introduction of change is eased if the symbols of change present no apparent alteration or modification of the culture's widely held symbols. . . .

One resolution of this difficulty . . . is to conduct a project with mixed teams of investigators, some of whom have instrumental skills, and others who have institutional skills. . . . Many organizations use similar devices. . . . For example, often an organization has two leaders instead of one. . . . The role played by one leader is to institute instrumental change. The role played by the other leader is to represent the institutional requirements of the organization. . . . When the instrumental leader causes friction, the institutional leader smooths it over or rephrases the requirements in more acceptable, warmer, more congenial, or more orthodox terms. It matters little who is the obvious leader, so long as the two cooperate (and can stand each other)![20]

In this connection, Hare cites with approval an illustration taken from a description of intelligence operations: a prisoner is interrogated by two men in turn who exhibit contrasting styles. The first is brutal, cruel, a martinet and threatens imminent death; finally, he leaves in pretended exasperation; the second then appears, who is kindly, sympathetic, offers the prisoner food and cigarettes and seems distressed with his plight. Finally, the confused and exhausted prisoner tells his secrets to the sympathetic interrogator.

Little more can be said here on this topic, but the administrator of systems analysis projects who seeks to blend instrumental and institutional skills for a given study can benefit from further study of the dual role approach to organization.[21]

Systems men, understandably, exhibit some concern with the problems of convincing clients that the systems sold to them are what they want, meet their purposes, and work as promised. But the missionary works of systems analysts, which radiate an Utopian glow of the promises of systems theory, do not often include the failures or diasters produced by them.

In principle, we have the technological capability of adequately feeding, sheltering, and clothing every inhabitant of the world. . . . In principle, we have the technological capability of providing sufficient education for every inhabitant of the world for him to enjoy a mature intellectual life. . . . In principle, we have the technological capability of outlawing warfare and of instituting social sanctions that will prevent the outbreak of illegal war. . . . In principle we have the capability of creating in all societies a freedom of opinion and a freedom of action that will minimize the illegitimate constraints imposed by the society on the individual. In principle, we have the capability of developing new technologies that will release new sources of energy and power to take care of physical and economic emergencies throughout the world. . . . In principle, we have the capability of organizing the societies of the world today to bring into existence well-developed plans for solving the problems of poverty, health, education, war and human freedom and the development of new resources. . . . If the human being has the capability of doing all these things, why doesn't he do it? . . . The answer is that we are not organized to do so.[22]

So begins a work introducing the systems approach to the layman. In view of such glowing promises, which suffuse the systems literature, it is appropriate for us to examine some of the many instances in which the systems approach has proven unsuccessful.[23] Ida R. Hoos' *Systems Analysis in Public Policy* is the first full-length attempt to examine systems analysis as applied to government and social problems from a critical perspective. She describes systems analysis in its beginnings in operations research and in the aerospace industry, traces its movement into the Department of Defense under Robert McNamara, and from there to its adoption by President Johnson for all departments of the federal government. But even within its own field, military hardware, systems analysis cannot claim unqualified success. Some military systems are primary examples of miscalculation and mismanagement. SAGE (Semi-automatic Ground Environment), a system for air defense, slipped schedules, exceeded cost estimates, and proved obsolete before completed and paid for. BMEWS (Ballistic Missile Early Warning System) was engineered to detect incoming ballistic missiles through electronic sensing systems. But BMEWS was so sensitive that it picked up energy reflections from the moon, flights of birds, and other nonballistic objects as though they were missiles. Only a lack of confidence in the system prevented counterstrikes.[24]

Researchers in the field have issued warnings against the dangers of reliance upon erratic and oversensitive systems. Moving beyond the purely military, systems analysis takes on slightly different forms when applied to government. Although the Department of Defense has been declared by congressional committees to be loose and flagrantly negligent, wasteful of defense dollars, a paradigm of waste and inefficiency, its touted systems methods of management have spread to other government agencies, under the name PPBS (Planned Programming and Budgeting System), which President Johnson forced all federal agencies to adopt in 1967. Along with this development went the diffusion of Defense Department systems experts into highly placed positions in government and into influential advisory oppositions. Since then it has spread to state and municipal governments throughout the United States and with it the notion that management techniques can be applied to all facets of public planning. Related to this is the rise of the "social indicators" movement.[25] Its literature is filled with references to social indicators, partial system models, which bring economic, social, political, and cultural variables together into a general systems theory. But social indicators are simply old-fashioned statistics in a new dress, similar to the old and familiar economic index numbers, such as the cost-of-living index, whose virtues as well as limitations, are well known. But these "objective measures" of the public good are relative, value laden, culturally defined, and subjectively conceptualized and are therefore simply unattainable. "This being the case, what may pass as and be accepted for social indicators may be so prejudiced, so biased, and so contrived as to be downright dangerous.[26] Hoos refers to Daniel Moynihan's judgment of this phenomenon as a McNamara colonization of domestic departments of government. Despite the shortcomings, gaps, and failings, these approaches have been sold and oversold and have become a integral part of the operation of the Department of Health, Education and Welfare. Related to the social indicators movement is the document published by HEW, *Toward a Social Report*.[27].

Hoos indicates that the kind of thinking found in *Toward a Social Report* is far inferior to that of even the social pathologists of half a century ago.

They, too, focused on crime, poverty, and health, and in terms no less normative. When they raised value-laden questions, however, they recognized that they could not expect definitive answers; underlying the current advocacy of social indices there is the insidious lure of techniques attainable but not yet harnessed. As for methods used in amassing the data on which the Social Report was built, there is no evidence that progress has been made since Durkheim. In fact, nothing present . . . rivals his insights . . . *Toward A Social Report*, itself a poor indicator of the state-of-the-art in sociology, stands as a persuasive argument against artifically contrived and inappropriate measures of the human and social condition.[28]

Hoos documents the spread of systems analysis activity throughout states and cities, under the support of federal agencies such as HEW, OEO, NIMH, HUD, into fields such as planning and policies, personnel management, health and hospitals, taxation, crime and corrections, pollution control, parks and recreation, social security, welfare, antipoverty programs. She finds that systems studies have become a secretive game of high-level government officials; failures and disasters are hidden or explained away on grounds of lack of time and funds, implying requests for still more funds. Finished studies are sequestered private property. Systems analysts, members of a tight community, are careful to whitewash colleagues' errors, and no politician will commit political suicide with an admission of error in having a systems study done.

In New York City, for example, tight security was imposed to keep a $500,000 NYC-RAND Institute report from public view. When the *New York Times* obtained a draft of the first volume and published an article on it, the city's Housing and Development Administration undertook an investigation to identify and bring criminal charges against the person who supplied the document. Only "public papers" must, by New York State law, be open to inspection; "investigations" and "studies" are exempt by New York City's Charter. Ten City Councilmen had earlier instituted legal action to force the city to make public the controversial housing report which, they claimed, had already circulated in private banking circles, and of which the chairman of the housing committee had not even been given a copy.[29]

The advantages of burying systems studies are obvious: Consultants and contractors have time to move on to other triumphs before the birds come home to roost in the form of a growing realization of the worthlessness of the studies; just as important, public officials gain the advantage of decision making that is not accountable to the public; thus, a hothouse environment for systems techniques is guaranteed. Hoos makes a point on the way such studies are sequestered and never used; again citing New York City

. . . where $75 million was paid to outside consultants in 1969, it was discovered, late in the game, that ten different studies of one bridge had been made since 1948, that the current Transportation Administration was unaware of six of them, and that not one of them was implementable or implemented. A sequence of administrations had reaped the benefits of public display of concern over the situation without making a further attempt at its amelioration. But even more important, if action *had* been taken, its results could, if politically strategic, have been blamed on outsiders. Far from insensitive to the many fruitful dimensions to be derived from systems work, the City of New York bought the entire back cover of an issue of the Institute of Management Science's journal (July, 1970) to broadcast "an opportunity to be at the center of complex urban systems analyses." The advertisement, unwonted in nature, called for any number of "talented individuals" with experience in "problem formulation, modeling, computer systems design and programming" to work at salaries from

$15,000 to $25,000. Proclaimed in the text was the intelligence, somewhat counter to current disclosures on the subject, that ". . . the mayoral staff of the Lindsay Administration is creatively utilizing quantitative analyses and computer technology with support from NYC-RAND Institute and consulting services such as McKinsey & Company. But all the problems are not solved and we need more talented individuals capable of project leadership and analysis."[30]

With New York City as a model, the smaller cities of the nation are not far behind. But a new problem emerges. When the Defense Department contracts for weaponry, the details of the system, the specifications for construction, performance, and evaluation, are all clearly understood by all sides. Further, the qualifications of the contracting firms are clear and tangible. But what happens when such contracts are let for work on civil systems in which intangibles prevail? Hoos cites a series of requests for proposals offered by the State of California, asking the aerospace industry to propose systems designs for waste management, transportation, care of the mentally ill, and for the "treatment" of criminals. Anatomizing the documents that flowed in from the systems firms, she found public relations techniques, and little else, prevalent. Thus of the approximately 50 proposals submitted to the state, most dwelt long on the successes of these firms, including their many defense and aerospace contracts. Among them was Lockheed Missiles & Space Company. "Lockheed's claims to managerial prowess, like those of its prime customer, the Pentagon, have been subjected to congressional scrutiny and found severely lacking in substantiation . . . the company has been accused of gross mismanagement of its own affairs." Lockheed wanted a $250 million government loan to save it from bankruptcy; the congressional committee stipulated that the request would be considered only if the chairman of the board and all its members resigned.

Even rarer and more noteworthy is the criticism of systems analysis contracts on the federal level. The Comptroller General of the United States reviewed the administration of three research research study contracts totaling about $600,000 awarded to the Hudson Institute by the Office of Civil Defense, of the Department of the Army, in 1968. The results were reported to Congress as a warning to exercise more careful control over contractors to obtain truly useful reports. The effort was estimated to have cost the government about $45,000 to $50,000 per man-year of work in its execution. All three studies were found wanting; one was described as adding no new thoughts and yielding nothing that was not previously known; another was seen as rehash of old tired ideas; a third was described as lacking sufficient depth to justify distribution. Herman J. Kahn, director of the Hudson Institute, defended its practices, maintaining that the Institute engaged in rather speculative and improvisatory forms of thinking, which because of its very imaginativeness and adventurousness, should not be expected to have a high success rate. The Institute would accept no contracts that limited its creative freedom. The Comptroller General's Office noted that the Hudson Institute's progress reports were submitted too late to change the direction of the work even if the information provided had indicated that changes should be made. Progress reports that arrive in the hands of resource committees to late for usable suggestions re, of course, commonplace in bureaucratic research. Thus systems analyses proliferate, and perhaps their greatest protection is that they are neither feasible nor implementable. Nevertheless, they continue to be sold as a practical method.

SYSTEMS ANALYSES IN TRANSPORT, CRIMINAL JUSTICE, WELFARE, AND WASTE MANAGEMENT

The practical results of systems analyses when applied to public affairs have not been encouraging.

. . . the Bay Area Rapid Transit system . . . built lines through and to many places, so laid waste by the protracted and inept construction stages that it may be years before former use patterns will be restored, if ever. Terminals are out in the fields and way stations stand as sole occupants of once prosperous and now desolate neighborhood business sections.[31]

Systems studies in criminal justice stress law enforcement at the street level and seek to create ever more comprehensive information networks designed to keep tabs on persons designated as "potential offenders." The approach disregards matters such as the problem-ridden administration of justice, court calendars, social environment, police activities, and the functioning of penal institutions.

In terms of waste management, a similar comedy was played by the State of California and its contractors. In 1968 the State Department of Health contracted with Aerojet-General Corporation for $175,000 to do a cost-benefit model of solid waste management in the Fresno area, a model it was hoped would be usable elsewhere. The funding came in large part from HEW's Solid Wastes Program of the National Center for Urban and Industrial Health. The purpose was to develop an optimum system for the management of solid wastes and to develop a technology for the determination that would be applicable for other similar regions throughout California and the rest of the nation. More than 18 models were developed; costs in 1967 dollars were projected to the year 2000. The result was a comedy of bad and even fraudulent methodology. Since there were no firm output standards in this field, the firm set out to develop some. They decided to interview 39 persons employed in state and country health departments; using what they called a "forced decision-making method," they asked respondents to rate waste materials with reference to 13 categories of hazard, among them flies, water pollution, air pollution, rodents, odor; thus, about three persons were interviewed for each bad effect.

. . . the number assignments . . . although actually resulting in an ordinal scale, were henceforth treated and manipulated as though they had . . . a ratio scale. The "hard data" were nothing but a crystallizing of the hastily contrived catalogue . . . into arbitrary and overlapping categories through an artificial procedure in which three individuals . . . were taken to represent the total community attitude. This was patently a parody on public opinion polling, itself not an undisputed means of gathering data. The perpetration of this methodologically nonsensical approach to a serious problem apparently failed to recognize either the simplistic solutions they were inviting . . . or the invitation to Madison Avenue and public relations-style treatment of the matter, that is, to manipulate people's attitudes and ignore the problem.[32]

The consultant's rating of their own performance was highly laudatory; they had achieved, they said, a goal-selection process "far superior to that of individual judgment." Even better, they had achieved this by developing a computer program. The result appears to have satisfied the needs of systems designers rather than the needs of society. The technology to determine the "optimal system" for disposing of solid wastes was a (reverse) popularity poll of various types of solid waste.

The "target date" for the waste management system was the year 2000; Hoos remarks that the twenty-first century as the time for the functioning of their plans is popular among systems analysts in fields such as transportation, urban planning, land use, and so on. The reasons for the year 2000 are clear: A long time span postpones the day of reckoning and offers great opportunities for more and more data gathering, data processing, and the manufacturing of projections and extrapolations.

In welfare, the proliferation of cost-benefit studies, information systems studies, and systems of rational choices show an inherent contradiction: While the purpose is to get people off welfare, the means employed entrench and expand the system to run itself ever more elaborately and expensively. The basic stereotype of the welfare recipient in systems studies is that of the ne'er-do-well who could work if he wanted to. Ignored are the problems of job availability and the disabling problems of welfare recipients as well as the disruptive effects of increasing the labor supply and thereby forcing other marginally employed persons out of jobs and into welfare.

Hoos reports similar results in education. Systems studies of education are of two general types: (1) one approach, derived from economics, is based on the idea of planning education to meet the projected manpower needs of the economy; (2) a second approach consists of tinkering with educational procedures on the basis of the PPBS approach. During the 1967 session of the California State Legislature, an Advisory Commission on School District Budgeting was instituted; its mandate was to develop a program budgeting system based on the procedures of the Department of Defense. Funded by national sources, it relied on a flying squadron of "program budgeting experts." "Something over $500,000 was allocated for a fifteen-district, four-year pilot project. The time and the money have been spent and extensions of both sought, but workable procedures are not yet in sight."[33] The expensive model developed by the RAND Corporation for the Advisory Commission failed to include the crippling costs, direct and indirect, of maintaining vast data banks; it did not take into adequate account the data already being gathered within the California Department of Education.

Designed with the assistance of Aerojet-General Corporation engineers, the multimillion dollar electronic data-processing system already in operation stores vast quantities of information on every facet of the State's educational establishment. Besides keeping track of the children, their age, address, school location, grade, I.Q. and achievement records, and the like, it can report for any given day how many children were sent home for reasons of illness. This Sorcerer's Apprentice approach to statistics gathering had its beginnings in the need for data for Average Daily Attendance reporting and reimbursement but may have outrun its usefulness. Unresponsive to most user needs, it could not properly justify its expensive existence when subjected to official interrogation by a committee of the State Assembly.[34]

Cost models, despite their limitations, have been constructed by colleges, relating various disciplines to their expenses, such as faculty salaries, supporting staff, space occupied, and campus upkeep. On such bases, administrators decide to close down specific schools or departments. Thus a cost-model might show that high-energy physics, with few students per class; a high faculty-student ratio; great space needs for expensive computers, accelerators, laboratories, equipment show low "output" as measured by numbers of graduates and their employment during periods of retrenchment in science. But cost models do not answer the broader questions about the importance to the nation of scientific research or about the meaning or value of any other field of education threatened by such cost models. The University of California is typical of colleges in its use of cost models. There, they create a "student flow model," tallying applicants accepted, matriculated, graduated, dropping out. Thus the university's retention rate is calculated, disregarding the specific details of student decisions. A "faculty flow model" records the "economic" and "social" characteristics—however defined—of faculty members. Their use in plotting "optimal hiring strategy" is not encouraging, the problem, of course, being: "optimal" according to whose goals and aims? Similar results are found by Hoos in "optimum class scheduling models." Schools using federal funds for construction purposes, measuring "output" by the number of classrooms built, have been known to raise room counts by the use of folding screens and temporary partitions, reducing rooms to closet size. Similar considerations of "optimum plant utilization" led to the institution of the quarterly, four-term calendar, disliked by students and faculty alike, and now being abandoned.

Systems analysts see educational planning as a form of "policy intervention," or "manipulation of the future," another game played according to its own rules. The economist Jan Tinbergen has developed widely influential econometric models, describing educational policy in terms of manpower needs for future economic development.[35] Those who organize production should also organize education to provide trained manpower suitable for projected economic needs.

. . . whether made explicit or not, the notion of measuring the rewards of education through labor force participation underlies most of the cost-benefit literature. For economists, wages earned are the measure of success. The wedding of educational planning to manpower needs is an enticing one, for it offers the convenience of quantifiable inputs and outputs. The counting of students and teachers, the enumerating of jobs, the matching of one with the other, all have appeal because they lend themselves readily to 'systematic' treatment. They also carry certain assumptions: that manpower forecasting is an exact science; that the values of education lie vocationally in the after-school life; that education is a kind of escalator that rides upward from rags to riches; and that the student population (and their parents) will willingly comply with the dictates of a system that, like the Queen in *Alice in Wonderland*, sets one painter to paint some roses red and another to paint some white.[36]

The assumptions are of course shaky. Manpower forecasting is not an exact science and has had its share of disasters—for example, the unemployment of electronics engineers hired so enthusiastically 15 years ago who are now laid off by the thousands. But beyond this the job market is in great flux due to historical, social, political, and technological factors; any orderly tableau of inputs and outputs has only the most short-lived validity; deductions, therefore, about the educational process, which necessarily requires a longer time-span, are grossly inappropriate. The logic of measuring education by labor-force participation could have amusing consequences for feminists. Carried to its extreme, such a measure could bar young girls destined for marriage and motherhood from education, because cost-benefit analyses would indicate that the money spent on their education yields no measurable return.

Other assumptions are dubious. The one-way causality between education and income is unproven. The "poverty syndrome" commonly mentioned in the literature includes many deprivations in addition to lack of education. Medical and physical defects based on poverty and malnutrition require other remedies and even make class attendance difficult. Cost-benefit analyses of several educational programs mounted as part of the war on poverty revealed that poverty and education were not clearly linked and that dollar gains from special educational benefits were hard to find.[37]

The systems theorists at NASA and at the California Institute of Technology have seen education as a promising field for their wares, which include sophisticated electronic alarm systems for high schools, designed for riot prevention; a talking typewriter; computerized instruction devices; closed-circuit television; and programmed instruction courses. Research on some of this gadgetry is supported by the Carnegie Foundation and by the U.S. Office of Education. But research has indicated that claims for programmed learning and for computer-assisted instruction are not justified. Nevertheless, big business has "discovered" education; and IBM, Xerox, RCA, General Learning Corporation, Raytheon, Westinghouse Learning Corporation, the Systems Development Corporation, and Lockheed, among others are busy selling hardware and software. Students parrot words and pass tests for which they have been primed, and so the package develops its own self-awarded seal of approval. But, Hoos indicates, this socially isolated, dehumanized learning experience is quite possibly the worst possible approach to the needs of poor children.

Perhaps their very cultural disadvantage can be attributed to the fact that an electronic box has been their mentor since birth. Surveys have shown that children under the age of six spend more than fifty-four hours per week in front of the illuminated box, their viewing time in inverse proportion to family income. By the time they reach school, they will have sat watching television more hours than they will spend in classrooms during their entire academic career. . . .[38]

Such children are to be brought to school to sit in solitary confinement in a booth, where the recorded voice in the earphones will say, "Yes, that's correct." But educators and psychologists know that, more than anything else, these children need socializing learning experiences in the company of real live children and teachers. Such technology is known to violate basic psychological principles.

The record is similar in the field of public health. Once again, the State of California's Office of Health Care Services requested a proposal for a management information system, to which 18 firms, mainly in aerospace, consulting, and accounting, responded. Lockheed Missiles and Space Company won the contract. Lockheed's description, in Hoos' words, "typifies the technical approach that confuses management of an enterprise with management of its record-keeping."[39] Initiative for the study came not from the persons directly responsible for the already existing California program, but from "management people upstairs." The expectable result: a rigid, slow, cumbersome system, unsatisfactory to all concerned, clients, doctors, and the state; and medical services were cut to the bone.

Costs of health care can be measured in dollar amounts, but even here the possibilities of error, inaccuracy, miscalculations, and poor estimates are great. Add to this the lack of empirical data and, even more, differences in definitions of illnesses, misdiagnoses, differences in methodology leading to lack of comparability; all this leads to a chaos of specious thought and imagery. Complicating matters is the need to compute the indirect costs of illness, which include wage and tax revenues lost; the intangible psychic costs of disease, central to a humanist view, cannot be measured.

The benefits are measured in the same terms: Lives saved and disabilities averted are measured in terms of earning power alone. For the economist, the worth of a human life is measured in terms of averaged future earnings of a person in a particular age group. But such a measure as a basis for making health decisions exceeds "not only the economist's methodological competence but the moral and ethical wisdom of any mortal man." Examining a table of lifetime earnings,[40] Hoos shows that the life of a 20-year-old male is worth the lives of two babies if they are boys, and three if they are girls. Women's earnings place them in the category of second-class citizens with respect to health care as well as education; for the considerable proportion of women not in the labor force, housewives' services had to be estimated at the level of the average earnings of a domestic worker, with no credit allowed for longer hours, a seven-day workweek, and full service rendered the family and community as a whole. Despite the warnings that the measurable is not necessarily the important and that cost-benefit analysis is an intellectual gadgetry in which self-deception is inherent, cost-benefit studies proliferate in the field of health.[41]

Perhaps in response to such criticism, systems men have begun to take note of their failures. One recent work, *Politicians, Bureaucrats, and the Consultant—A Critique of Urban Problem Solving*, by Garry D. Brewer, is perhaps the first to tackle the problem head-on.[42] Brewer is described as a political scientist and as a member of the senior staff of the RAND Corporation. His book, a study of the disastrous experiences of two cities, San Francisco and Pittsburgh, is a representative document.

Brewer attempts to discover the reasons why systems analysis failed to provide usable results—in fact the results were disastrous—for urban planners and systems experts in San Francisco and Pittsburgh.[43]

Under the Community Renewal Program (CRP), created by an Act of Congress in 1959, grants were provided to local governments for the creation of urban renewal plans. The guiding "philosophy" provided by federal officials stressed the desirability of getting away from piecemeal unplanned renewal efforts—described by one such official as "projectitis"—in favor of more integrated broad-gauged approaches, which today would probably be called "ecological." The usual rhetoric of the quantitative disciplines of "operations research, decision theory, cost-benefit analysis, input-output studies, information theory, and simulation models as well as sociological and manpower analyses" were invoked.[44] Brewer assumes that social scientists are all of one mind about systems approaches, and considers planning to be solidly grounded in the social sciences.

SAN FRANCISCO

San Francisco's application for a planning grant was approved in October 1962, a 26-month program budgeted at just over $1 million, with a final report to be delivered on April 1, 1965. Computer simulation techniques were to be employed as the major approach.[45]

Familiar programmatic statements of goals, purposes, promises, and exaggerated claims about the capabilities of modern systems methods in planning the futures of American cities were made. The participants and expert advisers included, as usual, a stellar cast of systems experts, planners, administrators, academics, and social scientists, among them C. West Churchman, Martin Meyerson, Anselm Strauss, in addition to a roster of operations researchers, "urban systems analysts," and urban planners on the staff of Arthur D. Little, Inc.

In Pittsburgh, a similar operation developed and a grant of over $1 million dollars was also obtained; again with a stellar cast involving the University of Pittsburgh's Center for Regional Economic Studies, the CONSAD Research Corporation, and a technical advisory council.

A computerized simulation model to replicate the demand and supply activities of the houseing market was developed for the entire city. A vast number of variables were fed into the model, including census data on households, number of members per household, income, race and rent-paying ability. In addition, housing data were supplied, including house type, condition, location, demand for housing. A third body of data included the division of the city into neighborhoods, including information on the "amenity attributes" of each housing area and classifications of land units; the aging of housing stocks was simulated by the use of Markov chains, and a full catalogue of data on zoning, assessed values, and capital improvement costs, rental values, estimates of population gains based on birth, death, and migration rates were fed in. The variables were built into the model, the program was constructed, and a simulation run was attempted. Results were so unsatisfactory—they failed to approximate what was actually going on—that a second and third version of the model were hastily developed.

The San Francisco Department of City Planning published the results of a comparison of the model's prediction of new construction in the city with what actually occurred for three time periods—1961–1962, 1963–1964, and 1965–1966—which showed the model to be unacceptable. All participants judged the model to be useless; distortion was excessive; the input/output printouts were virtually unintelligible; many assumptions about the static character of the socioeconomic variables built into the model were unwarranted; the computer program was so rigid and expensive as to make almost impossible the alteration of components of the model and disregarded long-term structural changes in the demographic features of the city.

When interviewed about what the city got from the project, participants, planners and other experts were virtually unanimous: "nothing," "very little," "negligible benefits," "the city got about 3¢ on the dollar."[46] Brewer concluded that the magnitude of the problem was far in excess of expectations, that differences in viewpoints among key participants helped debilitate the research, and that the state of available data and of adequate theory to organize these data into meaningful configurations were "tested and found wanting."

PITTSBURGH

The details of the simulation model constructed for Pittsburgh differed considerably from those of the San Francisco model. The Pittsburgh model contained more than 20 submodels and subroutines, among them spatial allocation models; an interindustry input/output model; revenues and expenditure projections; data on total employment, income, labor force and population, commercial employment, and population spatial distributions. Results were no better than in San Francisco. Conclusions were described as wrong and faulty, outputs as unintelligible, and as in San Francisco, the variables built in were static in nature and hence unable to fit the dynamic changes observed.

Strangely, Brewer remarks that although the model was judged to be useless for the Pittsburgh city planners and was unacceptable as a policy-making or policy-assisting device, it had very promising educational applications. One of the program's builders indicated that he would use it for teaching and research at the University of Pennsylvania; another academic agreed that although the report was useless from a practical point of view, it was the "best no-cost professional experience one could have ever had"; and a third professor incorporated his revised submodel into the METRO urban game he used for teaching purposes at the University of Michigan.[47] Thus we may expect a generation of city planners and simulations experts who have been raised on unworkable simulation models.

The simulation model exhibited other vast defects, among them a sensitivity to very slight fluctuations in variable inputs, which caused large, unbelievable, and erratic jumps in crucial variable outputs, such as projected rates of business closures due to falling sales. Another problem involved the apparently overlooked fact in setting up the program that some simple cross tabulations ("disaggregations")—for example, households by income—would require half a million additional pieces of data. In simpler terms, the programmers appeared to neglect the enormous data requirements involved in cross-tabulating large numbers of variables. Further, although the full model was run using Allegheny County as the analytical context, the model could handle only 200 census tracts, while the county as a whole has nearly 500, and computer time needed to execute the model increases by the square of the number of areal tracts. Further distortion of data was necessary to accommodate machine requirements in the form of data simplified and reclassified several times over.

Computer runs were costing $70,000 to $100,000, with little to show; at this point the planning director abandoned the approach.

In both cities planning officials developed healthy skepticism regarding the reliability of consultants who, it was felt, pursued their own interests and enthusiasms, turning in reports of little use to officials and of little relevance to the problems originally posed, and who were able to decamp without bearing responsibility for their contributions or suggestions.

Brewer concludes.

The Pittsburgh Community Renewal Program reinforces all of the lessons learned in the San Francisco experience and provides us with insight into several other difficult problem areas. If the experience is even partially representative, the long-term prospects for the integration of the computer into the urban decision process are dismal indeed. The five summary assessments from the San Francisco case as to (1) the magnitude of the problem, (2) the extent of gaps in orientations and expectations, (3) the status and adequacy of theory and data, (4) the structure of incentive and motivation, (5) the extent of the "appraisal gap," *all* hold for Pittsburgh with only minor shadings of emphasis. In addition we might conclude that (6) selling poor, high-risk, or speculative research is counterproductive; (7) the university's lack of control over its various resources and the systems of incentive and motivation pertinent to its personnel reduce its effectiveness for purposes of integrating the computer into the urban decision process; and (8) at current levels of ignorance, the hostility and potential virulence of a given political context for social science research should not be underestimated.[48]

Brewer fails to perceive, it seems, that what he calls "hostility . . . for social science research" may be based not upon "current levels of ignorance" but rather on simple experience—that is, on the clear consciousness of having been taken by academic carpetbaggers. Brewer, in my opinion, fails to draw conclusions inherent in his own study. For him, the problems involved here include bigger and better programs, better and more data, more conscientious personnel, the certification of systems personnel, and in general a fuller panoply of professionalization. The thought that systems approaches are intrinsically inadequate or inappropriate to the problems they claim to deal with remains invisible. In fact, the three chapters of historical and empirical data on two community renewal programs are set within a total framework that operates to affirm the dignity, profundity, and professional rectitude of systems thinkers, by attributing defects and failures to individuals who operate with less than full professional dignity and responsibility.

But the question of the evaluation of systems theory remains: If the systems, when applied to social processes, do not work, or work to the detriment of the social conditions they were intended to improve, and the possibility of abandoning systems approaches is never considered, the problem of systems theory as a professional ideology emerges in full force. As Hoos indicated, when such failures occur, the systems men tell us that systems theory is good, but that the application was poor in that particular case. But such reasoning is false because systems procedures claim to be a set of techniques—applicable techniques at that. How else are they to be judged but in performance?

The second line of defense for the systems theorist may then be that systems techniques are still in their infancy and can be expected to improve. Perhaps, but in that case, at least two *caveats* should be offered: (1) We must take care that the "infancy" of systems theory, which excuses it from accountability, should not become its oldest or its only tradition; (2) until systems theory achieves its majority, or at least its adolescence, it might be reasonably expected to moderate its world-shaking or world-governing claims, and even, perhaps, to take payment only upon the delivery of usable goods. But the systems theorists continue to make vast claims, paying lip service to their failures, while going on as if the lip service has obliterated all objections. At this point the question of systems theory as socially "helpful" or "harmful" is replaced by the question of systems theory as professional ideology.

NOTES

1. Philip M. Morse, "The History and Development of Operations Research," in Grace J. Kelleher, Ed., *The Challenge to Systems Analysis*, Publications in Operations Research, Operations Research Society of America, Volume 20 (New York: Wiley, 1970), pp. 21ff. See also E. S. Quade, "Military Systems Analysis," in Stanford L. Optner, ed., *Systems Analysis* (Baltimore: Penguin, 1973) pp. 121–122.

2. Morse, *op. cit.*, p. 27.

3. C. W. Churchman, R. L. Ackoff, and E. L. Arnoff, *Introduction to Operations Research* (New York: Wiley, 1957), p. 1.

4. W. E. Duckworth, *A Guide to Operational Research* (London: Methuen, 1965), p. 8.

5. See Van Court Hare, *Systems Analysis: A Diagnostic Approach* (New York: Harcourt, Brace, Jovanovich, 1967), pp. 164–166 (with Foreword by W. Ross Ashby); Duckworth, *op. cit.*, pp. 27–33 and pp. 113–114. Linear programming is also described in the following chapter, in the section on economic models. See also Jay W. Forrester, *Principles of Systems*, 2nd ed. (Cambridge, Mass.: Wright-Allen, 1968), Chapters 1–3.

6. See Duckworth, *op. cit.*, pp. 40ff; K. D. Tocher, *The Art of Simulation* (London: English Universities Press, 1961), Chapter 1; Thomas H. Naylor, Joseph L. Balintfy, Donald S. Burdick, Kong Chu: *Computer Simulation Techniques* (New York: Wiley), pp. 5–7; G. T. Jones, *Simulation and Business Decisions* (Baltimore: Penguin, 1972).

7. Game Theory, though related, is a separate technique, and is discussed in Chapter 5. See Duckworth, *op. cit.*, pp. 85–89, 116 and 150–156.

8. See Ida R. Hoos, *Systems Analysis in Public Policy* (Berkeley and Los Angeles: University of California Press, 1962), p. 44.

9. Two standard introductions are A. Battersby, *Network Analysis for Planning and Scheduling* (New York: Macmillan, 1964); and Burton V. Dean, Ed. *Operations Research in Research and Development* (New York: Wiley, 1963).

10. Duckworth, *op. cit.*, p. 109.

11. Duckworth makes such claims while acknowledging that they have yet to be used. *Ibid.*, pp. 95–101.

12. *Ibid.*, p. 112.

13. Quade, *op. cit.*, p. 125.

14. Hoos *op. cit.*, p. 1; Van Court Hare, *op. cit.*, p. 1.

15. Hare, *op. cit.*, pp. 215–216.

16. *Ibid.*

17. Jerome Kanter, "Integrated Management Information and Control Systems," in Optner, *op. cit.*, p. 199. See also C. West Churchman, *The Systems Approach* (New York: Delta, 1968), Chapter 7, "Management Information Systems"; and Hoos *op. cit.*, Chapter 7, "Management Information Systems."

18. Hare, *op. cit.*, Chapter 13.

19. Allen Wheelis, *The Quest for Identity* (New York: Norton, 1958).

20. Hare, *op. cit.*, pp. 420–421 and 424–425.

21. *Ibid.*, p. 426.

22. C. West Churchman, *The Systems Approach* (New York: Dell, 1968), pp. 3–4.

23. The discussion here is only a preliminary one; for more documentation see the major study by Ida R. Hoos, *Systems Analysis in Public Policy—A Critique* (Berkeley: University of California Press, 1972); my present discussion closely follows that of Hoos.

24. *Ibid.*, p. 25.

25. Raymond A. Bauer, ed., *Social Indicators* (Cambridge: M.I.T. Press, 1967).

26. Hoos, *op. cit.*, p. 79.

27. *Toward a Social Report* (Washington, D.C.: U.S. Department of Health, Education and Welfare, January, 1969). The report shows no author, aside from HEW, but the front matter (p. vii) lists 41 members of the Panel on Social Indicators, among them Daniel Bell, Raymond Bauer, William G. Bowen, James Coleman, Otis Dudley Duncan, Martin Feldstein, Bertram Gross, Philip Hauser, Carl Kaysen, Samuel Lubell, Daniel Moynihan, Harvey Perloff, Neil Smelser, and Marvin E. Wolfgang.

28. Hoos, p. 81.

29. *Ibid.*, p. 109.

30. *Ibid.*, pp. 109–110.

31. *Ibid.*, p. 134.

32. *Ibid.*, pp. 140–141.

33. *Ibid.*, p. 150.

34. *Ibid.*, p. 152.

35. See Chapter 5 for further discussion.

36. Hoos, *op. cit.*, pp. 154–155.

37. Thomas Ribick, *Education and Poverty* (Washington, D.C., The Brookings Institution, 1968), as cited by Hoos, p. 156.

38. *Ibid.*, p. 167.

39. Dorothy P. Rice, *Estimating the Cost of Illness,* U.S. Department of Health, Education and Welfare, Public Health Service, Health Economics Series, No. 6, Public Health Service Publication No. 947–6 (Washington, D.C.; U.S. Government Printing Office, May, 1966), p. 3.

40. Hoos, *op. cit.*, p. 180. Chapter 7 of her work indicates briefly the failures of systems approaches in "management information systems," once known as electronic data processing. These data systems have gained acceptance and prestige far beyond their actual accomplishments in the business world, and surveys indicate a growing disenchantment over the lack of demonstrable results. There is no proof that computers have helped managers to make better decisions, nor can the heavy expenditures find justification through any effect on profits. Despite these warnings and complaints, management information systems spread, and are now endemic in large organizations.

41. Garry D. Brewer, *Politicians, Bureaucrats, and the Consultants—A Critique of Urban Problem Solving* (New York: Basic Books, 1973).

42. Brewer conducts his inquiry entirely within the framework of systems theory and takes it for granted. The failure of systems approaches in practice makes no impression on him, and he appears to feel that the only problem to be dealt with involves better data, better subroutines, more elaborate programs (i.e., an intensification of techniques and efforts that do not work). Of the 14 chapters in his book, the first nine are devoted to the description of systems philosophy and are purely formal and conceptual in character. Only Chapters 10, 11, and 12 offer some substantive matter.

43. *Ibid.*, p. 103.

44. Brewer describes Stanford Optner as instrumental in persuading the city to adopt simulation techniques in its renewal study (p. 105); Optner remains a prominent representative of the systems approach; see Optner, *op. cit.*

45. *Ibid.*, Chapter 12.

46. *Ibid.*, p. 190.

47. *Ibid.*, p. 215.

Systems Theory and Economics

Several streams of economic theory and technique—including linear programming, the input-output theory of Leontief, the game theories of von Neumann and Morgenstern, the mathematical models of economic processes developed by Frisch and Tinbergen, and the general theories of welfare economics—have contributed to the rise of systems theory.[1]

THE PHYSIOCRATS' TABLEAU ECONOMIQUE

In historical terms the Physiocrats led by Francois Quesnay (1694–1774) were the first to attempt to comprehend an entire economy in systematic terms, by construction of abstract models that illustrated the flow of commodities throughout the entire process of production and consumption in a manner similar to modern input-output analysis. The Physiocrats' model divided the economy into three components: the farming class, the landowner class, and a "sterile" class that included manufacturers. The farming class was seen as the producer of all wealth, the surpluses of which flowed to the landowners, and from them to the "sterile" sectors of the economy. Thus the basis of wealth was rent on land, and the surpluses acquired by the landowners and spent by them on consumption flowed through the entire economy. A portion of this was returned to the farming class, which by its work on the land continually renewed the basic wealth disposed of by the landowning class for the benefit of the entire economic system.[2]

Although the specific normative assumptions and class biases included in the Physiocrats' schema have been thoroughly criticized, Schumpeter makes the point that they are irrelevant to the basic importance of the *tableau*.[3] First, the *tableau* achieved an enormous simplification in reducing the millions of economic flows and relations occurring in an economy to comprehensible form; especially important in this respect was the concept of income formation. Second, the simplification thus achieved opened up "great possibilities for numerical theory." Quesnay did concern himself with statistical data and "actually tried to estimate the values of annual output and other aggregates. That is to say, he did genuine econometric work. This aspect, too, has acquired new actuality in our time through the great work of Leontief. . . ." Third and most important, the *tableau* "was the first method ever devised in order to convey an *explicit* conception of the nature of economic equilibrium. It would seem impossible to exaggerate the importance of this achievement if admiring disciples had not actually succeeded in doing so."[4] Economists in the eras following exhibited a growing awareness of its importance.

But the discovery was not fully made until Walras, whose system of equations, defining (static) equilibrium in a system of interdependent quantities, is the Magna Charta of economic theory. . . . Now Cantillon and Quesnay had this conception of the general interdependence of all sectors and all elements of the economic process in which . . . nothing stands alone and all things hang together. And their distinctive merit . . . was that . . . they made that conception explicit in a way of their own, namely, by the *tableau* method: while the idea of representing the pure logic of the economic process by a system of simultaneous equations was quite outside their range of vision, they represented it by a picture . . . it visualized the . . . economic process as a circuit flow that in each period returns upon itself.[5]

Thus, the notion of the systems theorists of a system in which "nothing stands alone and all things hang together" has roots as much in economic theory as in other disciplines. But from the initial stages of such a conception in the *tableau economique* to its emergence as an ideology in the modern sense, a long pathway was followed through the history of economic thought. The next full step was taken, as Schumpeter indicates, by Walras. His work on equilibrium theory is now part of the basic materials of economic theory.[6] The language of most economics texts is remarkably "cybernetic" in describing the concept of general equilibrium. Thus one text speaks of "changes in the economy as occurring in three stages: (1) the *impact* effect of some initial change, (2) the *spill out* of the effects of this change onto other sectors of the economy, and (3) the *feedback* of changes into the sector where the impact originally occurred—imagery quite reminiscent of systems thinking.[7] But such imagery is to be found in Walras' own writings and is implicit in the entire theoretical stream of positive economics, stretching from Walras and Pareto through John Hicks, Paul Samuelson, and Wassily Leontief.[8] Nevertheless, the emphases within this stream are not identical, and we must examine each in turn. There are three—linear programming, input-output analysis, and game theory.

LINEAR PROGRAMMING

As described in Chapter 4, linear programming is concerned with the problem of the maximization of a firm's profitability under specific constraints. The original form of the problem is that of choosing one out of a number of different ways to manufacture a given good in the way that minimizes costs or maximizes outputs and profit. Similarly, the problem of what mixture of foods to be selected for a cattle feed that meets all nutritional requirements at minimum costs is subject to linear programming methods. The techniques developed to solve these problems are purely mathematical in nature, closely related to, but not identical with, the techniques of matrix algebra. Linear programming is complicated by specific constraints, especially the constraint that values to be computed for a specific variable must be nonnegative (e.g., negative amounts of a specific nutrient will not occur), a constraint that eliminates many solutions and forces upon the linear programmer a number of trial-and-error approaches and a heavy dependence upon computers for iterative solutions.

Although a generation ago differential and integral calculus were the principal tools of the mathematical economist, they have been overshadowed by linear programming and its close relatives, input-output analysis and game theory. Perhaps the most important assumption on which the technique rests is that relationships between inputs and outputs are linear; and although some attempts at nonlinear programming have been developed, their applications have been limited so far,[9] as have, to a lesser extent, those of linear programming.

Nonlinear programming has been developed in recognition of the existence of monopolistic or semimonopolistic situations that violate the assumptions of perfect competition and, accordingly, the assumption that marginal utilities remain constant. This is one of the few concessions made by econometricians to the existence of institutional factors and their influence on economic processes.

The mathematical bases of linear programming include the important theorem (or axiom) that a single best solution to a linear problem does exist, under specified conditons—that is, when the number of nonzero variables to be determined equals the number of constraints on the solution. The one best solution could not always be specified, but could be approximated to any desired degree of accuracy.

Another mathematical result that linear programmers find important for economic theory is the "dual" concept: Every linear programming maximization problem is associated with a related minimization problem. The basic idea of duality theory was first suggested, according to Baumol, by John von Neumann.[10] Where a maximization problem involves finding the appropriate combinations of variables to maximize output, the corresponding dual will specify what costs are to be imputed to each of the firm's scarce materials that go into the input. Thus the solution of the resources allocation problem becomes the solution of the pricing problem.

The main import of linear programming, in addition to its consideration of the technicalities of the manipulation of equations, appears to be in its claim that the techniques of linear programming act to vindicate the classic formulations of general equilibrium theory. Thus the view, prominent since Walras, that in a state of general equilibrium the supply of every commodity is equal to the demand for it and that the price mechanism tends to restore general equilibrium whenever local disturbances lead to a departure from it, is supported by the proponents of linear programming, who claim that their results supply mathematical proofs that Walras was in no position to develop.[11] In this connection, Seligman remarks:

In essence, linear programming became a vindication of the theory of economic equilibrium under competitive conditions. Now, it is evident that it has been a useful tool for individual managements, as in the oil industry. But, insofar as general theory is concerned, programming, aside from a clarification of concepts, does not seem to have added anything new. One almost commits the fallacy of composition by leaping to the conclusion that programming techniques are analogues for the economy as a whole. The theorists appear to have forgotten institutional limitations, and as yet developments in nonlinear programming have not provided theorems that are operationally useful for the total economy. Theoretical conclusions for a *political* economy must be held in abeyance.[12]

LINEAR PROGRAMMING AND WELFARE ECONOMICS

Linear programmers have argued that their techniques make significant contributions to the problems of classical welfare theory. Welfare economics, in undertaking to make policy recommendations on the basis of economic theory, has in essence argued in favor of the "invisible hand" of the free competitive market. But linear programming has added nothing new to the argument except the substitution of its techniques for the earlier and less elaborate mathematical apparatus of Walras, Pigou, and other representatives of the neoclassical school. The primary question of welfare economics—how resources may be allocated in all sectors of a society for optimal satisfactions of all needs—is treated with a newer technical apparatus in order to arrive at the same results: Free pricing under conditions of pure competition lead to more efficient allocations of resources among industries and greater levels of consumer satisfaction than under any restrictive conditions such as tariffs, redistributive taxes, or a socialist economy operating under command conditions. To be made manageable, however, the analysis introduces a number of ideal elements. Thus the vast variety of goods and prices to be found in any real economy are set aside in favor of an assumed uniformity of prices and qualities; illustrations are based on highly simplified fictitious cases that are taken as models for the total economy; problems of initial inequalities of income are disregarded, as are the facts that not all firms act to maximize their profits and not all consumers always act to maximize their marginal utilities. And as indicated by critics such as Joan Robinson and John Kenneth Galbraith, hidden diseconomies generated by unequal buying powers and by monopolistic or semimonopolistic conditions are also disregarded. Furthermore, the argument is limited to the narrow sphere of the distribution of goods bought by households and disregards public expenditures, such as expenditures on education or on health services, which though not immediately measurable, can have beneficial long-term economic effects.[13]

Finally, the argument takes no account of the distribution of wealth and income between families. The concept of a Pareto optimum is defined purely in physical terms without regard to the human beings involved. When purchasing power is unequally distributed, a position on the production-possibility surface might be reached where some consumers were overeating and some starving. This would satisfy the Pareto condition perfectly, for the starving men could not get any more without at least one of the overeaters getting less.[14]

Neoclassical economists have responded to these criticisms, acknowledging that only if abilities and wealth

were originally distributed in an 'ethically optimal' manner—and kept so by nondistorting, nonmarket interventions—could even perfectly competitive pricing be counted on (a) to produce an efficient configuration of production . . . and (b) to give people what they really deem is best for them, in accordance with their dollar-votes that now reflect equally significant social utility. If laissez faire were abandoned in favor of an ethically proper distribution of wealth and opportunity, the perfect-competition equilibrium could be used to attain optimally efficient and equitable organization of society. . . . Admittedly, the demands of people in the marketplace sometimes do not reflect their true well-beings as these would be interpreted by even the most tolerant and individualistic of ethical observers. . . . Were Smith alive today, he would say, "People are entitled to make their own mistakes in many matters, but it is arrogant to think that anyone who is not a minor or a lunatic is in every respect a sovereign will. . . ." Furthermore, the answer goes, public scrutiny and control of monopolistic imperfections is certainly called for where appropriate, and finally, there does exist a "prima facie case for study to see whether zoning laws, taxes or subsidies, and government expenditure and regulation should be initiated in some degree."[15]

Nevertheless, as neoclassical economists argue, not even the most efficient of socialist computers could organize optimal price allocations within a socialist economy, and would be forced to resort, as in fact they have, to a market and profit-accounting system. Thus the argument remains unresolved. The failure to resolve the argument stems of course from normative assumptions that are not explicit in the mechanics or logics of the discussions. Linear programming has not resolved the problem. In Seligman's words:

> Programming was merely a computational technique for discovering the relevant points of efficiency, an objective of the free market as well. Moreover, it was argued, programming did not have to distinguish between individual firms, for under constant cost conditions for everyone, the problem was expressed for the entire economy. But like the older theory, this was an ideal statement and suffered similarly from the same limitations. Once dynamic elements and conditions of monopoly were introduced, other questions and doubts arose. Linear programming had to be well spiced with reservations. To take but one situation, activities that were influenced by diminishing returns were not handled satisfactorily at all. And increasing returns could virtually nullify a linear solution which recommended a process or activity less concentrated than was desirable.
>
> Upon close examination, there was little fundamentally startling about linear programming insofar as the substance of economic theory was concerned. Its theorems have provided newer, more elegant "proofs" of standard theory, but as Baumol remarked, the revolution has been more in methodology than in content. . . . Only when the whole economy is viewed as a firm, as in a socialist system, does linear programming offer important possibilities for achieving a social balance between production, technology, and prices. But the technique then would have to be advanced far beyond present developments in order to encompass the infinitely more complex conditions of a total society.[16]

Baumol, among others, remains optimistic about the uses of linear programming in welfare economics, and noted that it is now (1965) being used in promising ways in the "operations-research type of analysis of specific problems of government and as training material for operations researchers who can learn from the special concepts of welfare economics to avoid some frequently encountered analytic booby traps," among them the notion of trying to optimize results for individual departments of a firm which may prove less than optimal for the firm as a whole. Thus emphasis in welfare economics has moved from rather abstract considerations to "very applied work and concrete problems of day-to-day economic decision-making."[17] Nevertheless, where linear programming has proven inconclusive in its results, input-output theory and game theory have attempted to fill the gaps.

INPUT-OUTPUT THEORY

Input-output theory, largely the creation of Wassily Leontief, is an extension of classical equilibrium theory to the study of the interdependence of economic sectors in a society.[18] In effect the *tableau economique* of the Physiocrats has been resurrected in more complex form and is more fully saturated with empirical data than was the largely schematic *tableau*. The standard textbook description of input-output analysis proceeds with a highly simplified model of an economy; for the sake of simplicity, the economy is usually taken as a three-sector economy.[19]

Thus we consider an economy of three sectors, perhaps agriculture, mining, and services, or coal, steel, and railroads, labeled as we wish. An input-output table for such a three-sector economy might look something like this:

	Users of Output (in million $)				
	(1)	(2)	(3)	(4)	(5)
			Rail-	Household	Gross
	Steel	Coal	road	Consumption	Production
Sellers of Output					
Steel	$ 200	$ 200	$ 100	$500	$1000
Coal	400	100	300	200	1000
Railroad	200	500	100	200	1000
Primary inputs (labor)	200	200	500		
Total value of inputs	$1000	$1000	$1000	$900	$3000

Each sector shown in the table is at once a producer of output and a consumer of inputs. Reading along the rows, the value produced by each sector and consumed in the other sectors is given; thus steel produced $200 million for use within the steel sector, another $200 million for use by the coal industry, and $100 million for use by railroads, with $500 million for final demand (household consumption). Thus the total value of output produced by the steel industry was $1000 million, of which $500 million was used up as inputs within the three sectors, with $500 million left over for consumption.

Reading down a column, each sector is shown as a consumer; thus railroads consumed $100 million in steel, $300 million in coal, $200 million in rail services, and $500 million in labor, the other sectors are similar.

A miniature economy which shows the interrelation of each sector to all the others and to the "final demand," or household consumption, sector is presented in table form. The total economy is a $3000 million economy, of which $2100 million is used in the intermediate production phases, and $900 million is available for consumption. Thus the $900 million represents the capacity of the economy to produce goods for final demand and for further capital growth.

The importance of such a table is the bearing it has upon planning for further economic growth and for changes in outputs from specific sectors. Thus if the output of steel were to be increased by a specific amount, say $100 million, the table offers the potential of indicating how much more input into the steel sector from the other sectors would be necessary. Since the steel sector uses some steel, the output from this sector would have to rise by more than this amount. The problems can be solved through the use of matrix algebra. The amounts in the table above would be converted to proportion of total output, and these proportions would serve as the coefficients of three linear equations in three unknowns, with the appropriate final demand amounts as constants.

Thus the equations would read as follows:

$$S = 0.2S + 0.2C + 0.1R + 500$$
$$C = 0.4S + 0.1C + 0.3R + 200$$
$$R = 0.2S + 0.2C + 0.1R + 200$$

A fourth equation, involving labor inputs, would also be added:

$$L = 0.2S + 0.2C + 0.5R$$

Thus if the supply of labor in terms of man-hours is available to meet the needs, the goals can be met; otherwise, more modest targets will have to be substituted. From this point on, input-output analysis moves into matrix albegra, and the equations are solved by matrix inversion. Thus the analysis seeks to show what additional units of input from appropriate sectors are necessary to produce an extra unit of output from a specific sector for final consumption. This is in effect the basic problem of input-output economics.[20]

The obvious intent and implication of such a *tableau economique* is its intended utility as a tool of planning for the growth of an economy seen as a totality. Such a *tableau* requires the collection of an enormous amount of empirical data, not merely for steel, coal, and railroads, but for virtually every sector of an economy. Tables have been constructed for as many as 450 U.S. industries, and even this does not represent enough fine detail. Thus input-output economists have increasingly devoted themselves to collecting such data for more and more sectors of the economy. By 1963 input-output tables for more than 40 countries had been compiled.[21]

Leontief relates that World War II provided great stimulus for the application of input-output analysis. The need to coordinate the American industrial effort for war production became evident early. The need for more aluminum for aircraft production alone had implications for the generation of electric power, for the allocation of copper and other resources, which were quickly evident. Leontief had earlier prepared analyses of the U.S. economy for 1919 and 1929; in 1943 and 1944 he presented the government with input-output tables of the U.S. economy for the year 1939, "the first input-output table prepared under official government sponsorship." The techniques appeared promising enough, he indicates, to encourage both the Air Force and the Bureau of Labor Statistics after the war to construct still more elaborate tables, involving larger and larger areas of the economy.

Since that time many other countries have developed such tables for their economies, among them developed and undeveloped countries, free-enterprise and planned economies, and the European Common Market nations. Still more elaborate tables are planned for the American economy, which will include 500 to 600 sectors. Leontief likens such an instrument to the 200-inch telescope; it will enlarge the scope of economics enormously.[22]

Leontief has indicated the affinity of input-output economics to the concepts and procedures of operations research and to general system theory in the sense that things are seen in their interdependence.

Modern Operations Research, which in the United States and Western Europe is rapidly transforming the management practices of private business, is a direct descendant of the mathematical theories of the profit-maximizing behavior of an individual firm. . . .

The new approach to fundamental problems of economic planning was also born of pure theory—the theory of general equilibrium, or rather general interdependence—which aims at a concise description of the structure and the operation, not of a single firm, but of the national economy as a whole.

In terms of this theory the economic system can be viewed as a gigantic computing machine which tirelessly grinds out the solutions of an unending stream of quantitative problems: the problems of the optimal allocations of labor, capital, and natural resources, the proper balance between the rates of production and consumption of thousands of individual goods, the problems of a proper division of the stream of current output between consumption and investment, and many others.

Each of these problems can—in principle at least—be thought of as being represented by a system of equations. Under conditions of perfect competition an impersonal automatic computer—to which we usually refer as the economic system—has been solving these equations year after year, day after day, before the mathematical economists even thought of constructing their systems.

To explain the structure and the operations of this miraculous computer has been the principal task of the neoclassical general equilibrium theory. Pursuing the analogy still further, one can say that—according to the modern version of the general equilibrium theory—the competitive mechanism solves the equations fed into it through application of the so-called iterative methods, that is, the method of successive approximation. To verify—this interpretation . . . one actually could construct a simple model of that system and then insert the system of equations describing it into an electronic computer. The machine can be programmed to work out the numerical solution without any outside interference, through a sequence of successive approximations.[23]

The problems are not so simple. Input-output analysis in its first formulations had to accept a number of simplifying assumptions. One of these is the assumption of constant returns to scale; that is, an additional unit of input is in a constant and linear relation to the resulting units of output. Furthermore, it must assume that "in any productive process all inputs are employed in rigidly fixed proportions and the use of these inputs expands in proportion with the level of output."[24] In addition, each industrial sector is conceived as producing only one homogeneous output, with no commodities produced jointly. Moreover, the input-output model for an economy as described so far was criticized as static and closed. It described a situation at a given point in time, making no provision for changes in technology that would change the levels of efficiency in an industry, thereby upsetting the system of input-output equations, nor did it describe changes in final demand over time. Leontief therefore moved to a second phase of input-output analysis, in the form of an attempt at creating an open, dynamic model.

The closed system produced a unique solution for a set of equations. The system could be opened by disregarding one set of basic relationships in the system, with the result that a unique solution is no longer available. Thus if household demand is regarded as predetermined and external to the system, more than one combination of inputs and outputs in the rest of the economy could be computed to satisfy that demand. The dynamic model could also be adjusted to account for changes in productivity due to technological innovations and by allocating some outputs to increased capital or inventory. The mathematical techniques differ from those of the closed system primarily in the use of inequalities rather than closed equations. Thus excess amounts are assigned to overproduction; negative amounts indicate idle capacity.

The accuracy of fit of the dynamic input-output computations to economic reality remains uncertain. Seligman evaluates the technique as being of only moderate accuracy, and still unproven.

In actual practice, the dynamic model did not work out as had been hoped. An effort was made to retain fixed coefficients despite the introduction of time. . . . But not enough work has been done as yet to make a true dynamic model possible. Consequently, questions of accuracy and the relevance of the data used were bound to arise. Moreover, the actual impact of technological change became highly important. Tests of the predictive capacity of the input-output matrix have not been very conclusive. In some cases investment requirements were overestimated when compared with actual levels. . . . Empirical work on input-output analysis has been done in a number of countries. . . . Yet not all economists have been willing to accept input-output analysis as a viable way of studying the economic order. . . . The idea of fixed coefficients seemed especially unpalatable. Yet there was a way of treating the problem of varying production functions, if the Leontief system were modified with some of the later tools of linear programming. In the latter, the various levels of input utilization became variables, and criteria could be set down for selecting one method rather than another, whether these were based on minimum cost or maximum "welfare.' Viewed from this standpoint, the Leontief matrix became a special case in the more general theory of linear programming.[25]

In addition to objections to the restrictive assumptions described, there were other criticisms, the strongest of which, Seligman suggests, is that input-output tables are "little more than convenient classifications of past data," for unless new tables are produced from time to time, they go out of date. Nevertheless, many economists, among them Paul Samuelson, came to the defense of input-output analysis through the development of various mathematical ingenuities, and the technique appears to be firmly established in economic analysis, if somewhat overshadowed by linear programming.

As a basis for interindustry economic research, it is now an established tool of public policy planning agencies in many countries.

GAME THEORY

Still another theoretical construction for which global claims are made is games theory, a formal and mathematical system for analyzing conflicts and competition; its earliest theoretical applications, besides games, were in the field of economic theory; from there, it has spread to political science, where it finds many advocates.[26]

The classic document of game theory is the *Theory of Games and Economic Behavior,* of John von Neumann and Oskar Morgenstern,[27] first published in 1944. The book was received on its first appearance with great enthusiasm, praised as a classic that moved economics forward as almost no other work. "Ten more such books and the progress of economics is assured," was a typical response.[28] Since then the literature has expanded to large proportions, with contributions by political scientists, linear programmers, and mathematicians as well as economists.

Von Neumann and Morgenstern conceived of their work as economic theory with direct applications. Dissatisfied with the possibilities offered by use of the calculus in economic theory, they felt that economic theory and problems needed a new set of mathematical tools, drawn from set theory, linear geometry, group theory, and axiomatic logic. Accordingly, the purpose of the work was to replace the outmoded and inapplicable mathematical tools hitherto relied on by economists. The authors are forthright in their belief that their theories are applicable to economic problems. They hope to find a way to bring the theory of games into relationship with economic theory, primarily by finding elements common to both. Once this is done, the relationship will be seen as not artificial; it will make the theory of games and strategy into the appropriate instrument for a theory of economic behavior. They state further that they hope to establish satisfactorily that the problems of economic behavior are identical to the mathematical ideas of suitable games of strategy. They hope to establish this "after developing a few plausible schematizations." [29] The extent to which these 'few plausible schematizations" beg the question at issue is a central problem.

The Formal Concepts of Game Theory[30]

Competition between business firms may be taken as a game, based on estimates of the probability of a rival firm's making pricing and marketing decisions and based on the probable impact on the demand and competitive position of the firms. Similarly, military battles, election campaigns, as well as economic markets, are all structurally similar in that their outcomes are in part dependent upon the actions of other participants with opposing interests.

Von Neumann and Morgenstern began their analysis with certain assumptions about the rationality of those engaged in economic activity, especially the assumption that those engaged in such activity wish to maximize utility.

Their basic assumptions also include the notion that individuals have a clear and consistent idea of their preferences and that their hierarchy of preferences is operational at all times.

They then discuss briefly the idea of a "one-person" game, of man against nature, on the model of the Robinson Crusoe parable common in economics. The single person seeks to maximize his utility preferences in competing with nature. But such a "game" is of little interest from a social point of view. The authors have done the most original and successful work in the area of two-person games.

Zero-Sum Games of Perfect Information

A competitive game in which one player wins what the other loses is referred to as a zero-sum game. By "perfect information" is meant that both players have complete knowledge of the other's situation, as in chess, where both positions are constantly visible and unlike poker, in which one player cannot see the cards held by his rivals.

Strategy

Important also is the von Neumann-Morgenstern concept of a strategy.[31] If instead of making decisions move by move, one at a time, a player writes out a full statement of what steps he plans to take for every possible contingency, the statement is considered to be a strategy "in normal form," in constrast to the decisions taken one move at a time, which is a strategy in "extended" form. The mathematical logic of game theory is based on the formal properties of a strategy in normal form. A strategy is usually illustrated by a table of strategies and their outcomes. A typical game-theory table of strategies follows:

		Strategies for Player II			
		Y_1	Y_2	Y_3	Y_4
	X_1	W	D	L	L
Strategies for	X_2	D	W	W	L
Player I	X_3	L	L	W	W
	X_4	D	D	L	W

If player I chooses strategy X_1 while player II chooses Y_1, the result is a win for player I; if player II had chosen strategies Y_3 or Y_4, player I would lose; the combination X_1Y_2 results in a draw. The entries in the table need not be simply "win-lose-draw;" numbers can be entered, both positive and negative, standing for amounts won or lost, or for percent of a market captured by firm I or II, or for percent of the vote gained by party I or II, and so on.

Such tables are used by game theorists to describe strategies and to develop basic concepts.

Game theory is actuarial in character and is concerned with the long-term expectation of wins or losses over many plays. In the table above strategy X_2 appears to be indicated for player I because this row contains two wins against one loss and one draw; for player II, no one column is better than any other in terms of overall expectations—that is, wins and losses are evenly balanced in each of the four columns. Where outcome tables show inequalities of advantages between two players (firms, nations, armies, individuals), game theory develops a number of theorems. Perhaps the most important is von Neumann's minimax theorem: In every finite, two-person, zero-sum game, there is a value V (the average amount that player I can expect to win from player II) that can be assigned to it. Player I will not settle for less than this amount, and player II will not concede any more. Thus, games of this type will settle down to what is called an "equilibrium point" (that combination of strategies in which the stronger player seeks a "maximum" strategy—

that is, in which he maximizes the minimum gains allowed him by the weaker player—and in which the weaker player seeks a "minimax" strategy—that is, he seeks to minimize the maximum gains of the stronger player—and in which the two strategies coincide in the same cell of the table of strategies and their payoffs). Thus in the following table the entries are the percent of the market captured by firm I (with the rest going to firm II) for each strategy.

		Strategies for Firm II		
		1	2	3
Strategies for Firm I	1	80%+	25%*+	40%+
	2	10%*	12%	30%
	3	20%	14%*	65%+

The worst possible outcomes for firm I are: (1) capturing only 25% of the market if, having adopted strategy 1, firm II adopts strategy 2; (2) capturing only 10% if, having adopted strategy 2, firm 2 adopts strategy 1; (3) capturing only 14% of the market if, having adopted strategy 3, firm II adopts strategy 2. Thus, operating on the pessimistic assumption that firm II will always play its best strategy, the three worst outcomes are those starred. Of these, the best "bad" outcome is that of strategy 1; this decision, based on cautious pessimistic assumptions, is sometimes called a *maximum* strategy, in which firm I seeks to maximize its minimal payoffs.

Assuming similar thinking on the part of firm II (who fears that firm I will always play its best strategy), the worst possible outcomes for firm II are: (1) losing 80% of the market if, having adopted strategy I, firm I adopts strategy 1; (2) losing 25% of the market if, having adopted its strategy 2, firm I chooses strategy 1; (3) losing 65% of the market, if having adopted its strategy 3, firm I chooses strategy 3. These are marked +. Since firm II seeks to minimize its maximum losses, its minimax strategy would be to play strategy 2, in which its worst possible loss of market is 25%.   The table illustrates the concept of an equilibrium point: If both players assume the maximum pessimistic and cautious point of view, their optimal strategies will tend to coincide; the maximum point for firm I is to play strategy 1, and the corresponding point for firm II is strategy 2; both strategies coincide at the payoff point of 25%.[32]

Thus if payoff matrices each have an equilibrium point, certain properties are deduced: (1) Each such game can be given a definite value—that is, the maximum amount that one side can expect to win or lose, given optimal play by both sides; (2) each side will have an optimal strategy that coincides at the equilibrium point; (3) matrices possessing more than one equilibrium point will demonstrate the property that these points are equivalent and, therefore, that choice between them and their corresponding strategies is a matter of indifferences; (4) a maximum strategy for one side may be disadvantageous if met by nonminimax strategies.[33]

At this stage of game theory the thinking employed is mathematical logic and mathematical induction. The nature of the thinking is exemplified by the following basic theorem, stated informally by Morton Davis:

> . . . every two-person, finite, zero-sum game of perfect information is strictly determined. . . . This means that one of the players has a winning strategy . . . which . . . will guarantee him a win, no matter what his opponent does. . . . We will show indirectly that one of the players has a winning strategy. That is, we will show that the assumption that neither player has a winning strategy leads to a contradiction. We will not actually detail a winning strategy; that is a practical impossibility in a game as complicated as chess. We will merely prove the strategy's existence. . . .

> We start by assuming that the game is *not* strictly determined. That is, whatever strategy A uses, there is at least one strategy B can select which will lead to a win for B. And whatever strategy B uses, there is a strategy A can pick which will allow A to win. In other words, the initial position is not a win for either player. Two conclusions immediately follow.

> 1. *Any position to which A can move from the initial position will not be a winning position for player A.* If this were not so—if A could go to some position X which was a winning position for himself, he would have had a winning position from the beginning. His strategy would be to go to X and then use the winning strategy that exists at this position.

> 2. *There is at least one position to which A can go from the initial position which is not a winning position for player B.* If this were not true, B would have a win from the start. He need only wait for A to make his first move and then play the winning strategy that exists at his winning position.

> Now we are almost done. From 1 and 2 it follows that there is a position immediately after the initial position which is not a win for either player. Call that position Y. In exactly the same way, we can show that there is a position immediately after Y at which neither player has a win. We can continue in this way indefinitely, and so the game can never end. (Once the game ends, one of the players must have reached a winning position.) This contradicts the assumption that the game is finite, and the theorem is proven.[34]

Game theorists acknowledge the inapplicability of such considerations to situations in the real world. Very few economic situations are finite, strictly determined, and offer perfect information. Rather, they regard the zero-sum, two-person finite game of perfect information as a first step.

UTILITY THEORY

The finite games described above are often too trivial to engage the competitive "instincts" of many players. People often fail to act to maximize their winnings. Lotteries, for example, have a negative value or expectation, but nevertheless attract large numbers of players. Thus game theory is forced to recognize the existence of psychological factors, which have to be mathematicized.

The point may be illustrated by considering two zero-sum games having the same average payoffs. In game one there is a 50-50 chance of winning or losing $10 dollars, in game two a 50-50 chance of winning or losing $10,000. Since the long-term expectations are the same for both games, they are mathematically indifferent. Yet an individual might well be indifferent (willing to play) to the first game and so fearful of losing the second that he or she avoids it, despite the equal possibility of a win. Game theorists are then forced to distinguish between expected payoff values and "utilities," which are subjective valuations of payoffs. At this point game theory becomes psychologized; individuals must establish weights for their preferences, and these weights must be constant and consistent. The illustration most widely offered is that of asking the person to establish his preferences. If he or she prefers an apple to an orange, and an orange to a pear, utility weights can be assigned to each of these if they are offered as prizes in a lottery with known probabilities. Given the preferences shown, he must prefer an apple to a pear; but if he prefers the pear, utility functions cannot be constructed. The functions are constructed by asking persons to indicate their preferences and the conditions under which they are indifferent to specific outcomes.

Thus individuals are assumed to be consistent and constant in their preferences; another assumption is that they will always gamble. Of course there are conditions under which persons might not choose to play: In a lottery in which there is a probability of .8 of winning $1, a probability of .1 of winning $1 million, and a probability of .1 of being executed, some individuals might choose not to participate. But game theorists assume that the player will always participate if the odds are good enough. Von Neumann and Morgenstern seek to construct measures that take into account the psychological weights assigned to various outcomes by the participants, which differ from the purely actuarial values of the expected outcomes. The psychological premises include the assumptions that items are fully comparable, that preferences are consistent, that players will prefer one lottery to another if it offers a higher probability of success, and that players are indifferent to gambling (i.e., if a person is offered a lottery ticket whose prizes include other lottery tickets, his attitude toward the compound lottery will be the same as if he had made all the probability computations, because he will be concerned only with the ultimate outcomes). Game theory, then is forced to base itself on debatable psychological premises.

TWO-PERSON, NON-ZERO SUM GAMES

Game theory has achieved its most decisive results in analyzing the structure of the two-person, zero-sum game of perfect information, partly because the determinate structure of the game makes it amenable to *a priori*, mathematical treatment. Its achievements in more complex situations have been much more tentative, and the absence of empirical reference has proven to be much more of a problem.

A non-zero-sum game is one in which both opposing sides can achieve positive gains, in part dependent upon how the other side plays.

The table[35] illustrates a basic feature of the non-zero-sum game. If both players adopt A as their strategy (AA) or both adopt B (BB), they get nothing. If player I adopts A and player II adopts B (AB), player I gains 10 units and player II gains 5; similarly, the joint strategy BA nets 5 units for player I and 10 units for player II.

If the players do not communicate, each may choose to adopt an optimal strategy; if player I chooses to play his strategy A ⅓ of the time and player II does the equivalent, both are guaranteed an average payoff of 10/13. But if they can get together and agree on choices of strategies, they can increase their payoffs. If players are not permitted to communicate, they can signal their intentions. Thus player I might choose the inferior strategy B, therefore offering player II the possibility of choosing A (payoff is then 5/10; if player II reads him correctly, on the next round they will arrive at the joint strategy AB, with payoff of (10/5); thus, they learn to avoid the zero-payoff cells of the table and to alternate on the cells having joint payoffs.

If players are allowed to communicate directly, they need not adopt the

		Strategies for Player II	
		A	B
Strategies for	A	(0,0)	(10,5)
Player I	B	(5,10)	(0,0)

choice of strategies as a channel of communication; then game theory is complicated by the possibilities of bargaining, collusion, threats, and counter-offers. At this point game theorists must rely upon practical experiments using real persons as players in order to arrive at some notions of how players will in fact behave. The primary difficulty with such procedures is that the games and the stakes involved tend to be trivial and, therefore, are doubtful models for reality involving favorite topics of game theorists such as price wars, military threats, election strategies, or advertising campaigns. The one important economic application of the theory of two-person nonzero-sum games occurs in situations of duopoly—that is, in industries dominated by two major firms. The logical strategy for duopolists, recommended by von Neumann and Morgenstern, is to collude on price fixing and market allocations, thereby avoiding costly price wars and the opening up of too many outlets.

Game theorists have elaborated nonzero-sum games to cover situations in which the payoff matrices show unequal payoffs. In such instances the stronger party can demand side payments in return for collaborating; it has been considered applicable to arbitration schemes, bargaining behavior, competitive bidding, nuclear strategy, and disarmament. But its applicability to economics proper is highly limited and depends upon artificial parlor-game experiments, to which must be added ad hoc theories about personality types, such as maximizers, rivalists, and cooperators. Its primary importance has been its adoption in political science as a model for international relations. But its value appears to be chiefly as a theoretical device; applications are few and limited. Furthermore, game theory is not unique in that maximal strategies for many payoff matrices can be derived using the methods of linear programming. Game theory becomes a special case of the latter.

MANY-PERSON GAMES

Von Neumann and Morgenstern also attempted to explore the structure of games involving three or more competitors. Since players can maximize their chances by forming coalitions, their discussion is primarily an analysis of which coalitions are likely to occur. As might be expected, those yielding maximum (Pareto-optimal) rewards to all concerned are favored; players will arrive at them by a mixture of bargaining and logic. Once the coalition has formed, the theory of many-person games collapses to that of the two-person game, the "person" in this instance being a coalition. Coalition formation and the bargaining around possible coalitions have been the object of elaborate though somewhat abstract theoretical developments. But again, these have had little or no application in economics.

Observers have, for the most part, expressed disappointment in the value of game theory for economics.

. . . the 13 years that have elapsed since the publication of *The Theory of Games* have seen no important applications of game theory to concrete economic problems. The theory of games has had a profound impact on statistics and on military science; in economics it is still merely a promising and suggestive approach.[36]

. . . game theory does need to account more effectively for the sort of communication which itself alters the condition of play: this is the dynamic problem. Perhaps the greatest weakness in game theory . . . is its basic assumption that a utility scale can be constructed. This more than anything else has threatened the applicability of its logic to the analysis of human conflict. But even more significant has been the extreme formalism of the theory, tending to give it an aura of aridity. It has threatened to become mere mathematical dexterity and technique. Game theorists have not yet been able to place their analysis into a broader framework.[37]

Game theory, like linear programming and other techniques, stands as a "promising start" that has provided a new framework for discussion, a new rhetoric of concepts, but whose concrete applicability and results remain in doubt.

DECISION THEORY

Game theory has had an important impact in a related field, statistical decision theory. The logical difference between the two fields can be described simply as follows: In game theory *risks* are undertaken with known probabilities for specific outcomes; in decision theory decisions are undertaken under conditions in which outcomes cannot be predicted in probabilistic terms. Thus it is describable as a "game" in which there is no way to estimate what the other "player" is likely to do. Decision theory coalesces with the newly emergent Bayesian statistics, which is beginning to establish itself with "classical" statistics (mostly created by Ronald Fisher, A. Wald, and J. Neyman), but which is by no means dominant or without resistance among statisticians.

Baumol indicates:

These considerations have, as yet, had little influence on the methods of applied statistics. Rather, they have affected mostly the relatively abstract and philosophical discussions of the foundations of statistics. In part, this is because decision theory is still in a rudimentary state and can offer no firm and final answers to the questions of statistical design.[38]

Game theory, then, like the other disciplines surveyed, has been seen as a promising new theoretical approach, but one whose empirical fruitfulness is not fully demonstrated. Nevertheless, like these other fields, its framework, rhetoric, and terminology have been adopted in other social science and administrative fields, perhaps without full comprehension of their strengths and limitations.

We turn now to the consideration of the migration into the social sciences—of images, language, and terminology drawn from the systems fields examined.

NOTES

1. Sources consulted include: R. Dorfman, P. Samuelson, and R. Solow, *Linear Programming and Economic Analysis* (New York: McGraw-Hill, 1958); William J. Baumol, *Economic Theory and Operations Analysis*, 2nd ed. (Englewood Cliffs, N.J.: Prentice Hall, 1965), J. von Neumann and O. Morgenstern, *Theory of Games and Economic Behavior*, 3rd ed. (New York: Wiley, 1953); Wassily W. Leontief, *The Structure of American Economy, 1919–1935*, 2nd ed. (New York: Oxford University Press, 1951); Morton Davis, *Games Theory* (New York: Basic Books, 1972).

2. Joseph A. Schumpeter, *History of Economic Analysis* (New York: Oxford University Press, 1954), pp. 239ff.

3. For criticisms, see Schumpeter, *op. cit.*, pp. 240ff; Joan Robinson and John Eatwell, *An Introduction to Modern Economics* (London: McGraw-Hill, 1973), pp. 8–10; Robert Heibroner, *The Worldly Philosophers*, rev. ed. (New York: Simon and Schuster, 1961), pp. 34–36.

4. Schumpeter, *op. cit.*, pp. 41–242.

5. *Ibid.*, pp. 242–243.

6. Ben B. Seligman, *Main Currents in Modern Economics*, Vol. 2, *The Reaffirmation of Tradition* (New York: Free Press, 1962), reprint (Chicago: Quadrangle, 1971), Ch. 5; see also Schumpeter, *op. cit.*, pp. 999ff.

7. Richard G. Lipsey and Peter O. Steiner: *Economics*, 3rd ed. (New York: Harper & Row, 1972), pp. 405–410.

8. Seligman, *op. cit.*, p. xi.

9. Baumol, *op. cit.*, Ch. 7; and Dorfman, Samuelson, and Solow, *op. cit.*, Ch. 8, provide descriptions of nonlinear programming techniques.

10. Baumol, *op. cit.*, pp. 103ff; and Dorfman, Samuelson, and Solow, *op. cit.*, Ch. 3.

11. Dorfman, Samuelson, and Solow, *op. cit.*, p. 381 and all of Ch. 13; see also Baumol, *op. cit.*, Ch. 7.

12. Seligman, *op. cit.*, p. 783.

13. Robinson and Eatwell, *op. cit.*, pp. 214–215.

14. *Ibid.*, p. 216.

15. Paul Samuelson, *Economics*, 8th ed. (New York: McGraw-Hill, 1970) pp. 610–611.

16. Seligman, *op. cit.*, pp. 783–784.

17. Baumol, *op. cit.*, p. 384.

18. Wessily Leontief, *Input-Output Economics* (New York: Oxford University Press, 1966), p. 134.

19. Leontief also proceeds in this manner; see *ibid.*, pp. 134ff. Other but essentially identical presentations are to be found in Baumol, *op. cit.*; Dorfman, Samuelson, and Solow, *op. cit.*; Samuelson, *op. cit.*; and Lipsey and Steiner, *op. cit.*

20. Baumol, *op. cit.*, p. 483.

21. W. Leontief, "Input-Output Analysis" (1965), Ch. 6 of his *Input-Output Economics*, p. 152.

22. W. Leontief, "The structure of the U. S. Economy (1965)," Ch. 8 of *Input-Output Economics*, pp. 157–158, 164–166.

23. W. Leontief, "Modern Techniques for Economic Planning and Projection" (English transla tion of *"Techniche moderne per la pianificazione e la privisione economica," La Scuola in Azione*, Ente Nazionale Idrocarbuir—ENI Scuola Enrico Mattei, Anno di Studi 1964–1964, No. 23), reprinted as: Ch. 20 of W. Leontief, *Essays in Economics—Theories and Theorizing* (New York: Oxford University Press, 1966), pp. 237–238, and as *Essays in Economics*, Vol I, 2nd edition (New York: International Arts & Sciences Press, 1976).

24. Baumol, *op. cit.*, pp. 480–481.

25. Seligman, *op. cit.*, pp. 439–440.

26. This chapter describes game theory only in its economic implications; Chapter 7, on the social sciences, discusses its migration to political science.

27. John von Neumann and Oskar Morgenstern, *Theory of Games and Economic Behavior* (Princeton: Princeton University Press, 1944); 2nd ed., 1946; 3rd ed., 1953; reprint (New York: Wiley, 1967).

28. Seligman, *op. cit.*, p. 773.

29. The discussion of game theory presented here does not claim to be exhaustive or even complete. It is intended simply to indicate ways in which games theory has been adapted to systems approaches, without exploring those ways in all their ramifications. Specifically, game theory has also been incorporated into social science thinking in nonsystemic usages. A full discussion of this development is, however, beyond the scope of the present study. For further discussion of this matter, see, among other sources: Stanford M. Lyman and Marvin B. Scott, *A Sociology of the Absurd* (Pacific Palisades, Calif.: Goodyear, 1970); reprint (New York; Appleton-Century-Crofts, 1970). Chs. 2 and 3 provide thorough surveys of this topic.

30. Von Neumann and Morgenstern, *Theory of Games and Economic Behavior*, 3rd ed. (New York: Wiley, 1967), pp. 1–2.

31. *Ibid.*, p. 79.

32. Baumol, *op. cit.*, pp. 530ff; von Neumann and Morgenstern, *op. cit.*, pp. 93ff.

33. Morton D. Davis, *Games Theory—a Nontechnical Introduction* (New York, Basic Books, 1970); reprint (New York: Harper & Row, 1973), pp. 16–18.

34. Following Davis, *op. cit.*, p. 67ff.

35. Dorfman, Samuelson, and Solow, *op. cit.*, p. 445.

36. Seligman, *op. cit.*, p. 779.

37. Baumol, *op. cit.*, p. 568.

The Societal Claims
of the Systems
Thinkers

The Convergence of Systems Thinking

From the beginning systems theorists have shown an ecumenical bent, claiming significance for their work beyond the limits of any one specific discipline. But only comparatively recently have synthesizing works appeared, designed specifically to establish the cross-disciplinary significance of systems concepts. (Earlier missionary writings were designed primarily to explain the nature of a special field.) In addition, meetings, conferences, societies, and yearbooks have been established, whose principal feature is their multi- or even pandisciplinary scope. Beyond this, a number of works of a different character have appeared. These syncretist works are designed to establish a philosophical framework that demonstrates the underlying unity of systems fields, and within which new world views can unfold. These works make impressive claims, among them the ability to resolve the philosophical dilemmas around old questions such as monism versus dualism, fact versus value, the meaning and nature of causality, the paradoxes of modern physics, and above all, to find the thread leading out of the labyrinth of epistemology. This chapter examines works of the systems thinkers that attempt a philosophical synthesis in both wider scope and closer detail than heretofore described.

BERTALANFFY'S GENERAL SYSTEM THEORY

General System Theory—Foundations, Development, Applications, published near the end of von Bertalanffy's career (1968), is his major attempt at a philosophical synthesis. He indicates that systems theory is a development

. . . which far transcends technological problems and demands, a reorientation that has become necessary in science in general and in the gamut of disciplines from physics and biology to the behavioral and social sciences and to philosophy. It is operative, with varying degrees of success and exactitude, in various realms, and heralds a new world view of considerable impact. The student in "systems science" receives a technical training which makes systems theory—originally intended to overcome current overspecialization—into another of the hundreds of academic specialties. Moreover, systems science, centered in computer technology, cybernetics, automation and systems engineering, appears to make the systems idea another—and indeed the ultimate—technique to shape man and society ever more into the "megamachine."[1]

From this starting point, von Bertalanffy moves to a general survey of biology, physics, psychology, linguistics, the social sciences, and history, describing systems trends within each field, but without doing detailed philosophical work. In this sense his work is, like much systems writing, primarily programmatic, selective, eclectic, and conceptual. The discussion and resolution of specific, substantive problems is simply not to be found. Nevertheless, the work is of symptomatic significance in that these traits set the tone for the systems literature that followed, not only among his disciples, but among independent workers who proved to be late converts. He appears to have left to his followers the task of creating the philosophical framework that would unify the systems disciplines.

ERVIN LASZLO'S PHILOSOPHICAL SYNTHESIS

Ervin Laszlo, a contemporary philosopher of systems theory, appears to be the leading representative of systems thinking in philosophic terms.[2] Laszlo's systems philosophy may be examined in his ambitious *Introduction to Systems Philosophy* and also in his somewhat more popularized work *The Systems View of the World*. His express intent is to extend systems theory into a general systems philosophy, which will serve as an instrument to "polarize the contemporary theoretical scene as a magnet polarizes a field of charged particles: by ordering the formerly random segments into a meaningful pattern."[3]

BASIS FOR SYSTEMS PHILOSOPHY

Modern philosophy, Laszlo indicates, needs a return to synthetic philosophy as a corrective to its excessive emphasis on analytic philosophy, which has produced "increasing logic but decreasing substance."[4] Analytic philosophy has cut itself off from fresh empirical inputs and needs new channels by way of the synthesis of scientific information coming from nonphilosophical sources. Philosophers must step out of current modes of philosophizing and return to a synthetic, but carefully reasoned, philosophy.

There are both intrinsic and extrinsic reasons for such a philosophy. Analysis requires specialization into ever narrower and isolated fields of inquiry; but the world does not consist of isolated patches; the patches interact upon one another. The study of chemistry produces insecticides; insecticides produce ecological, economic, and political effect. We are "part of an interconnected system of nature, and unless informed 'generalists' make it their business to develop systematic theories of the patterns of interconnection, our short-range projects and limited controllabilities may lead us to our own destruction."[5] But since most contemporary Western philosophers remain content to deal with abstract conceptual and linguistic problems and presupposition, the theoretical scientist, the concerned humanist, and the educator must take up the task of dealing with the problem.

Laszlo indicates that there is still another crucial task awaiting synthetic philosophy—finding an answer to the meaning of life. The so-called "advanced" societies of the world, which have provided their populations with material well-being and have removed most of them from the grinding struggle for existence, find that they are trapped within an existential vacuum, without meaningful reasons for existence. Such a predicament generates "violence, anarchism, and political witchhunts—directed against mostly imaginary scapegoats ('capitalists' or 'communists' or merely 'administrators'), and which generate intense interest in traditional religion . . . as well as in Oriental religions and mysticism. . . ."[6] The demand for "seeing things whole" is itself a healthy reaction to the meaninglessness generated by overspecialized and overcompartmentalized research and analysis. Laszlo quotes approvingly the psychoanalyst Viktor Frankl in his protest against modern reductionism as a form of nihilism, nihilism disguised as "nothing-but-ness." Reductionism is the passion for tracing complex phenomena or processes to their smallest parts.

. . . reductionism generates a multiplicity of limited-range theories, each of which applies to a small domain of highly specific events, but says nothing about the rest. (And if the specialist does use his particular theory to explain events outside its scope, he becomes a *"terrible generalisateur,"* which, replacing the legendary "terrible simplificateur" is the mark of the generalising specialist.[7]   We must oppose the analytic, atomistic way of thinking, the *holistic* way of thinking. Citing Abraham Maslow, Laszlo suggests that "the holistic . . . is the sign of the healthy, self-actualizing person. Insistence on the atomistic mode is in itself a form of mild psychoneurosis." Even more, among contemporary scientists and philosophers, it is a defense used to wave away unwelcome questions.

But there are other, intrinsic reasons for a holistic philosophy. First, it is necessary to adopt certain primary presuppositions about the the world:

1. The world exists.
2. The world is, at least in some respects, intelligibly ordered (open to rational inquiry).

Once admitted, these presuppositions permit us to escape the epistemological bind in which so many modern philosophers remain and permit us to begin the "rational mapping of the empirical world;" theory construction can begin. But certain secondary presuppositions are also necessary:

1. The world is intelligibly ordered in special domains.
2. The world is intelligibly ordered as a whole.

Specialists tend to adopt the first, looking upon it as a "fact of nature"; generalists favor the second. The author quotes Whitehead to the effect that every proposition refers to a universe exhibiting some systematic metaphysical character. We may never know if either or both of these presuppositions are true, but they do permit the rational construction of theories and may be regarded as having equal heuristic potential.[8] But the contemporary emergence of general theory from the workshops of the cyberneticians, information and game theorists, and others, indicates that they can exhibit "general order where the classical concepts show only delimited special orders"; "their advantage over other concepts is that they are capable of remaining invariant where others encounter limits of applicability. . . . Systems concepts may thus be thought about in terms of general metalanguage of scientific discourse . . . general systems terms make it easier to recognize similarities that exist in systems of different types and levels."[9] Specialists operating at lower, delimited levels are unable to account for the concepts and properties that emerge at higher levels. Laszlo hopes that systems philosophy can generate a language that will unify disciplines now separated by specialized concepts and terms. He argues for systematic and constructive inquiries based "on the assumption of general order in nature." Such an assumption, and the results of so building, are "no worse, and possibly better, than the supposition of special orders." Man may not be the center of the universe, "nor is the universe built in his image, but he is a part of the overriding order which constitutes the universe."[9]

Systems philosophy, further, is the logical next step in the slow progress of the philosophy capable of performing this task, which began with Plato's philosophy of universals, the categories of Aristotle, the scholastic metaphysics of the Middle Ages, and "the modern process philosophies of Bergson, Lloyd Morgan, Samuel Alexander, and Alfred North Whitehead." Systems philosophy reintegrates "the concept of enduring universals with transient processes within a nonbifurcated, hierarchically differentiated realm of invariant *systems,* as the ultimate actualities of self-structuring nature. Its data come from the empirical sciences; its problems from the history of philosophy; and its concepts from modern systems research."[10]

The demand for such a philosophy is becoming recognized by progressive thinkers both in the Soviet world and in the Western world. But now, in addition to the new disciplines mentioned by von Bertalanffy (general system theory, cybernetics, information, decision, and game theories, and others), Laszlo now wishes to add to the list systems philosophy, the first detailed and self-critical formulation of the systems world view, "as the paradigm of general theory in contemporary thought."

THE METHOD OF SYSTEMS PHILOSOPHY

General systems theory represents a new world hypothesis; Stephen Pepper, who had, in 1942, described four "equally adequate" world hypotheses—mechanism, formism, organism, and contextualism—has recently added a fifth world hypothesis that he now regards as likely to be even more adequate than the others: "This is selectivism based on the 'root metaphor' of a purposive, self-regulating system."[11]

The philosophy to be developed is perspectival.[12] General systems synthesis will consist of the building of models of models.[13] The world hypothesis offered here is such a model, "mapping into potentially quantifiable constructs certain recurrent general features of the scientifically observable universe." Such models are constructed using two basic approaches: one approach observes the world "as we find it" and makes statements about observed regularities. The second approach, an axiomatic one, conceives of the set of all possible systems and seeks to reduce it to a more reasonable size. The first is Bertalanffy's approach; the second, Ashby's.[14] Although Ashby appears to believe that these are two sharply divided forms, they seldom appear unmixed with elements of the other. There can hardly be pure observation without some conceptualization; similarly, no axiomatic concepts exist uninfluenced by some empirical observation.

Laszlo now indicates a shift in the tenor of current scientific thinking, away from the empirical-imaginative method toward the deductively applicable axiomatic method. This shift, he says, was caused by the fate of Newtonian physics. Newton "created the impression that there were no assumptions in his theory which were not necessitated by observational and experimental data. Had this notion been correct, his theory would never have required modification: observation and experiment could not have contradicted it." But such contradiction occurred with the Michelson-Morley experiment of 1885, the experiments with black-body radiation some 10 years later, and subsequent work leading to Einstein's work. But citing Northrop's comments on Werner Heisenberg, Laszlo concurs that the theory of physics is not something based on description of experimental facts or deduced from them; rather, as Einstein has emphasized, physicists arrived at their theories by speculative means. Unfortunately, philosophers still think of physical theory in terms of induction followed by deduction, with the notable exceptions of a few such as Whitehead.

What is required is a form of creative deduction, a creative flight of imagination such as is to be found in Einstein's theory of relativity, Morgan's theory of genes, and Yukawa's theory of mesons. Systems philosophy will attempt a disciplined flight of the imagination, emphasizing the isomorphy of laws in different fields. Instead of focusing on first-order models of the world, as the sciences do, its data will be the theories constructed by the sciences; thus it will construct second-order models. "Its basic conceptual assumption is that the first-order models refer to some common underlying core termed 'reality,' and that this core is generally ordered." The special orders described by the specific sciences can be integrated into a scheme of general order.

HIERARCHY AS UNIVERSAL PRINCIPLE

Laszlo regards the concept of hierarchy as a universal principle, operating in all realms: inorganic nature, organic life, social life, and the cosmos. He indicates that, in his view, wherever development occurs, it takes hierarchical forms. That hierarchical systems develop more rapidly from their constituents than nonhierarchical ones has even been demonstrated mathematically; they "suffer" less when decomposed to simpler levels. Regardless of whether we speak of physical systems, living species, or social systems, "we find that those which are likely to be around" are organized on hierarchical lines. The others are not to be found in the records.[15]

The concept of hierarchy provides him with a framework within which to view the entire universe; at the base of the hierarchy is the space-time manifold; above this, in ascending order, he places fundamental energy-condensations; electrons, nucleons, photons, radition quanta; atoms. These constitute the lowest level of what he terms the macrohierarchy; above these on the scale are the various levels of the "microhierarchy" (terrestrial): molecules, crystals and colloids, cells and protoorganisms, organisms, sociosystems and ecosystems, culminating in the global system. Above the global system, the macrohierarchy resumes: stars and planets, stellar clusters, galaxies, galaxy aggregations, and the final culminating point the metagalaxy (the astronomical universe).[16] This scheme is diagramed in a large triangle within which are sets of similar triangles corresponding to the levels described; the microhierarchy, ranging from atoms through organisms and up to the global system, is a subtriangle embedded within the triangle representing the macrohierarchy.

Systems theory will serve not merely to explain phenomena within each of these levels; that, in fact, is the task of the specific sciences, within which systems approaches are being developed independently. Even more, the task of systems theory is that of coordinating these independent systems models in a general theory of systems. There are, of course, unsolved problems on all of these levels, as well as unsolved problems involved in the task of coordination.

General systems theory will proceed toward the solution of these problems by making various redefinitions; especially important in this respect is the shift in scientific thinking away from substantial entities toward relational entities, from "objects" to "field theories." Having classified the natural systems occurring on the various levels mentioned, it will then seek to map "invariances" by means of "creatively postulated" systems constructs. These are also referred to as "commonalities" that underly the manifest behavior of organized entities. These commonalities are "the general laws of natural organizations." [17]

With these preliminaries out of the way, Laszlo hopes to establish a world-encompassing systems philosophy. It is organized hierarchically, corresponding to the hierarchical order described above. To say that it is world encompassing is not an exaggeration. His program, using concepts of what he calls "systems-cybernetics" as a descriptive method, offers an ordered view of physical, biological, and social systems; then of cognitive systems and a theory of the mind. From this the study claims to create a framework for an ontology, a philosophy of nature, of mind, and of an epistemology, a philosophy of human freedom, a normative ethics, an ethos for a new age, and a new metaphysics.

Such ambitious claims deserve to be carefully examined; because of space limitations, I only briefly describe the materials offered as a new approach to physical and biological systems and devote more attention to his description of social systems and the framework for a normative ethics and new metaphysics—the latter two presumably having major social import.

THEORY OF NATURAL SYSTEMS

Laszlo regards physical, biological, and social systems as subclasses of natural systems as such, and all are governed by an overriding principle: *adaptive self-stabilization,* to be found at work among atoms, organisms, societies. Thus any fluctuating of a system will give rise to forces that bring it back to its stable configuration. More than this, such systems (i.e., systems with "calculable fixed forces")will reorganize their fixed forces and acquire new parameters in their stationary states when subjected to a physical constant in their environment. Here he incorporates Ashby's principle of self-organizing systems.[18] Natural systems in general go to ordered steady states. But since some of these are relatively unstable, the system will eventually reach states at which it sticks, these states being the more stable ones. This movement of a system toward its natural steady state is a *selection* process. Laszlo adopts Ashby's (widely quoted) example of a computer whose storage locations are filled with random digits, which are subject to a dynamic law, such that the digits are multiplied in pairs. But by the laws of arithmetic, even times even gives even, odd times even gives even, while only odd times odd gives odd. Thus sooner or later the system will "selectively evolve" toward the even digits. The process continues: Among the evens, zero times any number gives zero, so that ultimately the zeroes prevail. Thus, the system has evolved toward maximum resistance to change.

It is hard to understand how this illustration has come to be so widely accepted as a symbol (model?) of universal processes, but even harder to understand the conclusions drawn by Ashby and Laszlo: "Adaptive self-organization inevitably leads toward the known biological and psychological systems." He further quotes Ashby: "In any isolated system, life and intelligence inevitably develop; and: every isolated determinate system obeying unchanging laws will develop 'organisms' that are adapted to their 'environments.' "[19]

Neither Laszlo nor Ashby seems to find any difficulties in such statements; I may venture to suggest a few: (1) In view of the claim of systems theorists that everything belongs to the world system, can there truly be an "isolated system?" (2) It has not yet been reported that simple mechanical or electrical systems have shown signs of life or intelligence; surely the laboratories and test tubes would be full of them. (3) The phrase *isolated determinate systems* and the phrase *obeying unchanging laws* each begs a number of important questions. (4) Finally, since any "isolated system" that fails to develop life and intelligence would presumably not be a true "system," I suspect that a circularity or tautology is present: A "system," by definition, must evolve. How do we recognize a system when we encounter one? Because it evolves, shows steady states, shows adaptive equilibrium, and so on—that is, because it behaves like a system.

LASZLO'S SOCIOLOGY

The author's view of the social order is subsumed within his symbolic scheme for the entire universe, or world system. In describing this, he uses a formalistic, semimathematical symbolism, on several levels: (1) theory of natural systems; (2) theory of physical systems, (*a*) atoms, (*b*) biological systems (organisms), (c) social systems (human societies). The formal symbolism is as follows:

1. R = f(α, β, γ, δ), where α, β, γ, δ are independent variables having the joint function R (natural system).

2. α: conative relation of parts resulting in *ordered wholeness* in the state of the system ("systemic state property").

3. β: function of adaptation to environmental disturbances resulting in the *reestablishment* of a previous steady state in the system (system cybernetics I).

4. γ: function of adaptation to environmental disturbances resulting in the *reorganization* of the system's state, involving, with a high degree of probability, an overall *gain* in the system's *negentropy* and *information-content* (system cybernetics II).

5. δ: *dual* functional-structural *adaptation:* with respect to subsystems (adaptation as systemic *whole*) and suprasystems (adaptation as coacting *part*) (holon property).[20]

Atoms show systemic order, as does the table of elements. The forces acting as constraints within the structure of the atom are systemic forces. Atoms can even be described as exhibiting adaptive self-stablization under conditions of electron bombardment and adaptive self-organization, as in the processes of building up more complex elements, which go on within the interiors of stars. He even speaks of the heavier elements as "evolving." Next, the holon properties of atoms are revealed in the hierarchy of shells within atoms (by *holon* is meant any entity that is a whole in regard to some components and a part in relation to others). The intersystemic hierarchies of elements are illustrated by molecule formations.

Biological systems are next described under the same scheme; these two show wholeness, order, self-regulation (biocybernetics), adaptive self-stabilization (homeostasis produced by regulative glands, metabolic adjustments, etc.), and even a more advanced form of biocybernetics, in their built-in capacities to learn. Learning itself takes various forms: There is nonconscious learning (visceral learning), habituation, trial-and-error learning, latent learning (trial-and-error learning unmotitived by specific problems or frustrations), and insight learning. Along with perception and innate release mechanisms, all these forms of learning contribute to intelligence, the most highly developed form of adaptation. Biocybernetics operates on still another level, that of evolution. Short-term adaptations (bio-cybernetics in short-term range) are subsumed under the heading of *learning;* long-term adaptations under the heading of phylogenetic evolution. Thus, a cycle unfolds of self-stabilization, external forcings, leading to internal constraints, leading to adaptive self-organization on a higher level, with self-stabilization returning again only to be succeeded by new forcings, new internal constraints, and a search for self-stabilization on still higher levels.

The organic realm shows both intrasystemic and intersystemic hierarchy, the first of course revealed by functional physiology, the second by network of ecological and sociological systems. At this point social systems are described. Social philosophers, in the speculations on the nature of society, conceived of schemes such as the social contract or the abandonment of the struggle of each against all in favor of a ruler (Hobbes); but with the rise of sociological theory, society was recognized as an entity in itself.[21] The first such models— for example, that of Herbert Spencer—were organismic in character, but such views fell out of favor after Spencer's time. Current social theory is still divided on the issue of whether "social entities" are real "existents" or are "theoretical, constructed concepts;" hence, the division of contemporary sociology into functionalists and structuralists.[22] The sociologist's working tool, he tells us, is the theoretical system, but its validity must be determined in each instance.

The author defines sociology as a science that, like the other sciences, grasps pattern and order. Societies, like other systems, exhibit the same Properties: order, openness, adaptive self-stabilization, homeostasis, adjustment under feedback, and ultrastability. Societies evolve. For example, an agricultural society exhibits "pattern-maintenance"; but under structural change in the direction of increased differentiation, it transforms itself into an industrial society, and conserves the new pattern at a new level of differentiation.

But Laszlo draws his illustrations almost exclusively from sociologists who are already systems theorists: Talcott Parsons, Walter Buckley, S. N. Eisenstadt, and other avowed members of this school like Kenneth Boulding, Karl Deutsch, N. J. Demerath, and R. A. Peterson. The reader would be hard put to guess that there are sociologists who do not share the system view.

As though emphasizing the catholicity of the systems view of the world, he cites a Soviet theorists, E. Kol'man, who speaks of Communist society as a complex, open, dynamic system with ideal self-regulation, having at its disposal the feedback from numerous subsystems, and compensating "automatically" for deviations from the state of dynamic equilibrium.[23]

Even Max Weber appears to be a systems theorist:

The basic subsystems of the social system proper are its various institutions, constituting agencies of control of the behavior of individual members. Max Weber conceived of the distribution of institutional constraints in modern societies as the "bureaucratic machine" and emphasized its relative independence from the intentions and ideas of any group of individuals. The person engaged in the machine—the "bureaucrat"—is only a single cog in a functional mechanism which assigns to him a fixed set of routines. The official is entrusted with a set of tasks and cannot arrest and change the mechanism at will. Such directives or instructions can be issued only at the top, envincing the hierarchic nature of the mechanism. Moreover, such a mechanism cannot be destroyed by a group of persons opposed to it, such as anarchists, for it does not reside in public documents but is rooted in the orientation of man for placing himself, and obeying, a hierarchical set of constraints. (sic). Both the governing and the governed are thus conditioned. Thus, even with the formal structure of a bureaucratic mechanism is destroyed, it is reconstituted again due to the conditioning of most persons in the society. . . .

Bureaucratic mechanisms typify not only state administration, but all social institutions where concerted action is systematically carried out. The centralization of control functions, and the chain of command from the center (decision-making group) to the periphery (groups implementing the decision), is a general property of social institutions, whether we consider the legal, political, economic, or even the intellectual and cultural sphere. As a result the situation in which any given individual acts is composed of other individuals in ordered sets of relationship to him.[24]

Laszlo cites *From Max Weber* (Gerth and Mills) as the source in this discussion, but gives no page reference for the appropriate passages. But it would be a matter of some interest to know what Weber said about bureaucracy. I assume that the relevant passage in Weber is the following:

Once it is fully established, bureaucracy is among those social structures which are the hardest to destroy. Bureaucracy is *the* means of carrying "community action" over into rationally ordered "societal action." Therefore, as an instrument for "societalizing" relations of power, bureaucracy has been and is a power instrument of the first order—for the one who controls the bureaucratic apparatus.

Under otherwise equal conditions, a "societal action," which is methodically ordered and led, is superior to every resistance of "mass" or even "communal action." And where the bureaucratization of administration has been completely carried through, a form of power relations is established that is practically unshatterable.

The individual bureaucrat cannot squirm out of the apparatus in which he is harnessed. In contrast to the honorific or avocational "notable," the professional bureaucrat is chained to his activity by his entire material and ideal existence. In the great majority of cases, he is only a single cog in an ever-moving mechanism which prescribes to him an essentially fixed route of march. The official is entrusted with specialized tasks and normally the mechanism cannot be put into motion or arrested by him, but only from the very top. The individual bureaucrat is thus forged to the community of all the functionaries who are integrated into the mechanism. They have a common interest in seeing that the mechanism continues its functions and that the societally exercised authority carries on.

The ruled, for their part, cannot dispense with or replace the bureaucratic apparatus of authority once it exists. For this bureaucracy rests upon expert training, a functional specialization of work, and an attitude set for habitual and virtuoso-like mastery of single yet methodically integrated functions. If the official stops working, or if his work is forcefully interrupted, chaos results, and it is difficult to improvise replacements from among the governed who are fit to master such chaos. This holds for public administration as well as for private economic management. More and more the material fate of the masses depends upon the steady and correct functioning of the increasingly bureaucratic organizations of private capitalism. The idea of eliminating these organizations becomes more and more utopian.

The discipline of officialdom refers to the attitude-set of the official for precise obedience within his *habitual* activity, in public as well as in private organizations. The discipline increasingly becomes the basis of all order, however great the practical importance of administration of the basis of the filed documents may be. The naive idea of Bakuninism of destroying the basis of "acquired rights" and "domination" by destroying public documents overlooks the settled orientation of *man* for keeping to the habitual rules and regulations that continue to exist independently of the documents. Every reorganization of beaten or dissolved troops, as well as the restoration of administrative orders destroyed by revolt, panic, or other catastrophes, is realized by appealing to the trained orientation of obedient compliance to such orders. Such compliance has been conditioned into the officials, on the one hand, and, on the other hand, into the governed. If such an appeal is successful it brings, as it were, the disturbed mechanism into gear again.

The objective indispensability of the once-existing apparatus, with its peculiar, "impersonal" character, means that the mechanism—in contrast to feudal orders based on personal piety—is easily made to work for anybody who knows how to gain control over it. A rationally ordered system of officials continues to function smoothly after the enemy has occupied the area; he merely needs to change the top officials. This body of officials continues to operate because it is to the vital interest of everyone concerned, including above all the enemy.[25]

Max Weber cannot be represented as a systems theorist, or even as a "forerunner" of systems theory, and his discussion of bureaucracy carries an entirely different import than is suggested by Laszlo. First, Weber describes bureaucratic structures as purely contingent developments; there is nothing in either the structure of the universe or in the "essence" of the society that makes bureaucracy inevitable, or even desirable. Other forms of power structure are equally possible. Second, it is not the "orientation of man for placing himself, and obeying, a hierarchical set of constraints" that creates bureaucracy, but specific economic, social, and legal conditions. Third, Weber by no stretch of the imagination can be regarded as seeing bureaucracy as a desirable development; to the contrary, he expressed concern and even despair over the bureaucratization of society. Weber's views should be known to the reader, and to Professor Laszlo.

This passion for bureaucracy . . . is enough to drive one to despair. It is as if in politics the spectre of timidity . . . were to stand alone at the helm; as if we were deliberately to become men who need "order" and nothing but order, who become nervous and cowardly if for one moment this order wavers, and helpless if they are torn away from their total incorporation in it. That the world should know no men but these: it is in such an evolution that we are already caught up, and the great question is therefore not how we can promote and hasten it, but what we can oppose to this machinery in order to keep a portion of mankind free from this parcelling-out of the soul, from this supreme mastery of the bureaucratic way of life.[26]

Professor Laszlo's bland description of bureaucracy as occupying a "natural" place in the systems hierarchy, buttressed by quotations from other systems theorists such as Karl Deutsch, ignores the problematic and dysfunctional aspects of bureaucracy. His reification of hierarchy as a process linked to cosmic processes is, of course, important, but only in a symptomatic sense. That "sociocybernetics" is a doctrine tailor-made for bureaucratic elites receives no mention.

INTERSYSTEMIC HIERARCHY

There are more than intrasystemic hierarchies; Laszlo now foresees the development of intersystemic hierarchies

ranging from formally constituted multi-national states, such as the USSR, through economic blocs, such as the Common Market, to more loosely federated inter-governmental organizations . . . such as the OAS, its counterparts in Europe and Africa, and, at the most encompassing level, the struggling world-federation essayed in the United Nations. These supra-social (yet in themselves still *social*) systems carry forward the Chinese-box hierarchy of particular social systems in the inter-systemic direction, generally losing articulation proportionately to their hierarchical level. Here we are dealing with relatively new *international* systems, superimposed on the established national and international modules and still lacking the degree of differentiation and integration of the latter.[27]

Thus, he tells us, the world system is only at its inception; but "concerned scientists and humanists see the emergence of a concrete, self-maintaining global system as a precondition of human survival." In fact, the next 30 years may bring "the most crucial struggle of mankind's existence"—the struggle to create a global civilization as an organic whole. In the vanguard are such groups and individuals as The World Institute, the Society for General Systems Research, Marshall McLuhan, Buckminster Fuller, and Kenneth Boulding. That supranational systems are "emergent" in contemporary social interaction should be evident to all, he says. These form "systemic environments" within which smaller national and subnational systems will find stability and coordination. He expects that the emergence of a global system will bring new and unique properties.[28] The description of these emergent properties, to be sure, will have to await the event.

Laszlo fits the social world into his systems scheme only by treating that world in the most cursory way possible, and by evasion of the specific, the concrete, and the multidimensional variety of social forms. Objections to the systems view of society are often phrased in terms such as *social technology* or *technocracy*, and there is little here to contradict such objections. But a more serious objection is one that Laszlo himself might take seriously; he even refers to it. He quotes from S. A. Rich[29] that scientific method treats concepts *as if* they were real. This fictional process is justifiable if it leads to results. One may speak of a social group or "an ounce of ether," provided that something can be done with the idea.[30].

The ultimate question confronting systems theorists is what can be done with the idea, outside of exposition, definition, conceptualization, schematicization, diagramming, and programmatic declaration. What substantive theoretical or even "applied' sociological, political, economic, problems have been resolved by the elaboration of systems theory? We await the answer.

SYSTEMS THEORY AND VALUES

Having placed social systems within his framework, Professor Laszlo then moves on to other matters. He offers material on cognitive systems; a theory of mind; and in the ambitious second part of his book, systems frameworks for ontology, a philosophy of nature, of mind, for an epistemology, and a metaphysics. Most of the chapters are primarily philosophical in character and need not detain us. But three chapters do make social claims: Chapter 12 ("Freedom: Framework for a Philosophy of Man"); Chapter 13 ("Value: Framework for a Normative Ethics"); and Chapter 14 ("Survival: Framework for a New Age Ethos").

Chapter 12, on human freedom, can be discussed briefly. The argument proceeds within the limits of well-known philosophical discussions of human freedom. In the light of determinate causality, can human beings be free? That is, is speaking of freedom meaningful? Laszlo suggests that man as a physical system is determined, but when viewed as a psychophysical, natural-cognitive system, he is free. Although cognitive systems involve, as a first stage, the experience of percepts, and as percepts are determined in their character outside of the congitive system, the cognitive systems are not free on this level. That is, percepts and percept-patterns are determined. But a cognitive system is free to choose among percept-patterns and, further, is free to choose among constructs used to interpret and react to percept-patterns. A percept-pattern known as *water* may be interpreted according to three modes: (1) It might be interpreted from the common-sense point of view as something to drink, wash with, cook with; (2) it might be interpreted from in the scientific mode as having specific gravity, freezing point, boiling point, composed of H_2O; (3) it might be interpreted aesthetically in poetic or painterly terms. Thus the cognitive system appears to be capable of free choice.

Even further, natural systems possess freedom; they are free to choose the neural pathways by which stimuli come to them; even a limited brain-system contains enough cells and linkages between the cells to make possible an astronomical number of pathways by which any one set of stimuli can be routed. They also appear to possess freedom with respect to their choice of *conations* (response actions in the form of efferent signals fed to the cortical centers that control action). Laszlo here refers to Donald M. MacKay's notion of "logical interdeterminancy." An actor possesses the feeling (or illusion) of having freedom of choice; at the same time, an external brain analyst sees the actor's brain as fully determined. But if the external analyst conveys this information to the actor, this would change the brain state of the actor to another such state, and so is self-falsifying. For the rest, Laszlo's discussion merely recapitulates philosophical discussion in Kant, Descartes, and Whitehead and adds nothing to them. The third part of Laszlo's Chapter 12 amounts more to a plea for the notion of freedom than a demonstration of its existence or its desirability; these have in fact been assumed throughout. The best he can say is: "I believe that it can be shown that there is no necessary incompatibility between individual freedom and determinate social organization."[30] This cannot be proved or disproved on the basis of experience, but can be at least argued on the basis of "conceptual premises." The single individual, he tells us, is the apex of the organismic hierarchy.

A similar imagery is applied to man in the social order. Since atoms and cells possess "holon properties," there is no reason to believe that man or his social systems would constitute an exception. If man is a holon, "the conditions which satisfy his systemic requirements are compatible with those which render his social behavior functional and determinate in regard to his superordinate social system." Although there is no decisive empirical evidence one way or the other, the hypothesis, at least, can be maintained without contradictions.[31]

What is good for society, it seems, is good for the individual; the two are easily reconciled. We just haven't noticed this before.

In view of the grandiose claims made for systems thinking, one would have hoped that it would have provided a stronger basis for the value of freedom than these somewhat feeble pleas; the impression given is that although systems theory doesn't support the notion of freedom very strongly, it doesn't contradict it. In this case the preference for freedom is anchored not in systems theory, but in the values of the systems theorist or his readership. But then systems theory might provide the same anchorage for other values inimical to freedom. In fact its emphasis on hierarchy suggests this.

. . . communication in a differentiated technological society functions as a means of determining the decision-making capacity of individuals, rather than their immediate behavior. Behavior-determination is typical of insect societies and other multiorganic systems, where high degrees of plasticity and self-determination are ruled out by evolutionary and physiological factors; in *human* societies, behavior-determination is encountered only on the lowest levels of a hierarchy, where the learning-capacity of individuals is either at a minimum or has not had a chance to develop. (A case in point is the private in the army who is told not to think, but to obey.) However, decision-making determination is the rule on all medium and high strata of human societies, where the learning capacity of the individual role-carrier is a precondition of his effective functioning. Ideally, such determination is effected in the process of higher education, where not patterns of behavior, but patterns of information-handling and decision-making are to be taught. Having been "in-formed," the educated person is more determinate in many respects than the uneducated one: he perceives, and is able to respond to a wider range of situations: he can assert his individual judgment in regard to many more consciously held goals . . . Given our definition of individual freedom (as the self-determination of the given person by means of his historically evolved adaptive organization) social differentiation is not anathemic but propadeutic to individual freedom: it stimulates learning and therewith widens individual autonomy. . . . We thus get a parallel, mutually reinforcing evolution of differentiated complex systems on the individual and social levels. Highly differentiated modern societies tend to stimulate the learning capacities of their members, and highly learn individuals tend to create increasingly complex social roles.[32]

The sociologist's suspicion that systems theory is a self-serving doctrine for intellectuals is heightened by such phrases; this is a blueprint for what is widely termed *meritocracy:* freedom to command for those at the top of the hierarchy and freedom to obey for those locked into the system. Such authoritarian tendencies will not be checked either by the general imagery of systems theory or by the feeble plea to allow some slight leeway within the operation of the system—so-called freedom.

This issue raises a question not specifically dealt with by the author or by other systems theorists: What values will systems theory generate, or what values will it serve? Systems theory seems to resolve no problems connected with values. It has failed to "prove" that some values are "valid" and others not; in failing, it provides no solution for problems of social conflict and clashes. A theory that does not reconcile conflicting values fails to deliver in perhaps the most important area of social theory and thereby fails to live up to its grandiose claims. If it cannot resolve problems in this area, its claims for theoretical or even practical importance in social theory must be discounted. Furthermore, if systems theory in itself provides no support for some values against others, clearly it puts itself at the disposal of values originating outside of systems theory. It may therefore prove equally as useful to those who prefer domination, authority, and control as ultimate values as to those who sentiments favor "freedom," however defined. So far, then, systems philosophy provides no solution for value problems.

But perhaps we have judged Laszlo's presentation prematurely. In his Chapter 13, titled "Value: Framework for a Normative Ethics," perhaps we may hope to find a resolution to problems of value. One hardly has to state, perhaps, that the central problem for philosophers of value is the current view that values are not "facts," conflicting values cannot be resolved by scientific or philosophical methods, and that the anchorage of values in social contexts has led to a universal "relativism', about values, in view of the variety and contingency of social contexts which generate conflicting values. Systems theorists might do well to recall Weber's discussion of the matter, which one might have supposed could be taken for granted among social theorists.

"Scientific" pleading is meaningless in principle because the various value spheres of the world stand in irreconcilable conflict with each other. The elder Mill, whose philosophy I will not praise otherwise, was on this point right when he said: If one proceeds from pure experience, one arrives at polytheism. This is shallow in formulation and sounds paradoxical, and yet there is truth in it. If anything, we realize again today that something can be sacred not only in spite of its not being beautiful but rather because and in so far as it is not beautiful. You will find this documented in the fifty-third chapter of Isaiah and in the twenty-first Psalm. And, since Nietzsche, we realize that something can be beautiful, not only in spite of the aspect in which it is not good, but rather in that very aspect. . . . It is commonplace to observe that something may be true although it is not beautiful and not holy and not good. Indeed it may be true in precisely those aspects. But all these are only the most elementary cases of the struggle that the gods of the various orders and values are engaged in. I do not know how one might wish to decide "scientifically" the value of French and German culture; for here, too, different gods struggle with one another, now and for all times to come.[33]

So oriented, we may examine Professor Laszlo's chapter on normative ethics. After a brief mention of Plato's notion of the good as the highest form, the discussion moves to modern "metaethical" philosophy, whose analyses of words and concepts as used in everyday contexts are dismissed as supplying no solutions.

The classical moralists never derived moral prescriptions from goals and values; rather, they were the result of theoretical inquiries into the nature of man and his surrounding social and physical world. But we know that people often have highly mistaken ideas about themselves and their world; ordinary common-sense statements are no guide on these matters. Instead, we must rely on statements "based on careful methodological inquiry."

Here too, in ethical theory, the meaning of moral terms such as *the good, should, ought,* and the like can be derived from "informed scientific theories," and not from the unreliable materials of everyday usage.

Laszlo too appears to believe that a clue to the solution of "value problems" is given by natural science. Values have appeared to be arbitrary choices of human beings. But with the development of sciences such as cybernetics, systems theory, and information theory, mechanisms that operate on the basis of their own programmed goals can be described; these are values inherent in the systems themselves. Thus the fact-value gap is obsolete because all natural systems, Laszlo says, are themselves goal-directed systems.

A new literature on values has emerged from awareness of these developments, produced by men such as R. B. Perry, John Dewey, Rollo Handy, Abraham Maslow, Stephen Pepper, and Laszlo himself.[34] Thus the way is found for fusing facts and values, their fusion residing in the purposes and goals of human beings, rooted in the "telic nature" of physical, biological, and even artificial systems. Descriptions of the states of these systems will include concepts such as *goal, purpose, preference.* "The way is opened for a return to normative and naturalistic ethics, offering definitions of moral terms in light of invariant theoretical constructs. We shall review two examples of such ethics, both of which show marked similarities with Platonic moral theory and equally marked dissimilarities with contemporary schools of analytical meta-ethics."

One of these is Kenneth Boulding's correlation of goodness with higher and more complex levels of organization; the second is the biologist James Miller's identification of norms as the *strains* within an organism that correct toward a goal or norm. To these, Laszlo adds his systems definition of values: "a state of the system in which its percepts match its constructs is a value-state."[35] By this is meant:

If the cognitive system knows his own norms and reflects on the degree to which they are approximated in his actual situation, he correctly identifies those of his states as the most valued ("good") in which his actual perceptual experience is optimally matched by his construct sets . . . the match is a state brought about by the system through self-stabilizing and self-organizing activities representing an adaptation to the *factors* of change governing conditions in the environment. Thus a match does not ignore the future: it is the outcome of predictiveextrapolative adaption to the limits of the system's capacities.

The match is never absolute, but optimum. Similarly, "negative values" attach to "those states of the cognitive system wherein he is unable to meet the challenge of the environment." With sufficient information and energy-input, "the system is healthy and prospering: mentally he is characterized by cognitive satisfactions and physically by the fulfillment of energetic requirements. He attains a state of optimum value in fulfilling biological, psychological, and socio-cultural needs."[36]

This material is offered seriously. But a few objections may be hesitantly raised, though they are no doubt obvious to the reader. First, the problem of value conflicts is totally evaded. "Natural systems," each embodying values, have been known to clash; one may ask how systems theory resolves the questions of whose value was the "right" one. So far it does not. Further, even individual systems have been known to carry (or at least to profess) values that are self-contradictory. People have been known to want both their penny and the cake, and indeed both have "value." Second, the "genetic programming' or "instinctual programming" of individuals, by whatever name, is clearly not deterministic of all values. Are political affiliations genetically programmed? Do religious conversions (perhaps to be renamed "cosmic value reprogramming") result from neural surgery? Here, Laszlo's suggestion that adaptation is the ultimate yardstick of values could be interpreted to mean that the highest value is found in joining the winning side. Third, the concept of adaptation is of no use in judging the aspirations of individuals; which aspirations can, before the event, be judged as "unrealistic," and which can be judged as within the potentials of the "natural system?" Laszlo's conception of value as a match between percepts and constructs extrapolated into the future appears to glorify timely adjustment to trends, and no more. And his image of the "good" state of a "cognitive system" as well-fed, well-informed, with, presumably, a portfolio of futures in a rising market, is astonishing. Nature is, everyone knows, filled with strife and predation as well as love and cooperation; all of these are acted out somewhere and sometime by "natural systems," often by the same ones. But what this means for the basing of a civilization on value-systems is not even touched. The purpose of civilization, as Nietzsche pointed out long ago, is to be something other than nature. One might add, especially so, since one can find almost anything in "nature." But all this is elementary; what is of interest here is that material of this type is seriously offered. This is a theoretical and practical problem to the solution of which this study hopes to contribute.

Laszlo, to be fair, is aware of strife and conflicts, but his discussion cannot be called adequate. He describes these problems under the heading of "value misidentifications"; it is because some systems have reflective consciousness that errors can occur. After all, perfectly valid goal-strivings might go astray through misunderstanding either of the environment or of the goals themselves. These errors can take two forms: One is that individuals will forget that they are parts of higher level systems and will act egoistically, antisocially, pathologically. The second form is somewhat surprising: "On the opposite side of the coin, value-misidentification take the form of martyrdom and unreasonable heroism in the service of a social cause. To die for one's tribe or country when one's death is not a precondition of the survival of that collective entity is just as much a misidentification of intrinsically healthy goal orientations as to refuse to serve it." I should have thought that the "opposite side of the coin" could be described a little differently: the overemphasis on collective values by which individuals are stunted; the use of collective patterns, values, and instruments for the impoverishment and stultification of some groups and the enrichment or glorification of others, in either material or ideal ways, or both. But to wash out of this discussion the vast problem of conflicting world views, as they are embodied in societies and social groups, as though these are merely problems of cognition and definition, cannot be taken seriously; once again, it is of symptomatic interest only.

UNIVERSAL VALUES

Nevertheless, the author assures us that the solution is in sight in the form of a universal value system. The cultural relativism of the early twentieth century is now superseded; by doing cross-cultural studies, universal values common to all cultures can be located, both by theory construction and by empirical findings. Behind the apparently endless variety of cultures is a fundamental uniformity.[37]

Thus every society shows mutual obligations between parents and children; all societies disapprove of suffering, killing, stealing within the in-group; reciprocity is also essential. These, according to Kluckhohn, are "universal categories of culture." Firth too, he says, finds universal values in all cultures: morality, courage, incest taboos, and so on. The discovery of these universal social values, Laszlo tells us, shows that men are, like all natural systems, holons.

The ultimate value, the author tells us, is founded in the struggle of systems to maintain themselves against fantastic odds within their environments. By their very struggles, such systems indicate their embodiment of the value-proposition: Existence is good, based objectively on the "total complex of mental, physical, and societal states which add up to the attainment of bodily and cognitive goals."

AN ETHOS FOR A NEW AGE

The social ethos of the new age will be based on the ideal of reverence for natural systems. Man has for too long attempted to subdue and exploit nature; what follows is a description of world pollution, an evocation of the term *ecology,* and the belief that ideological differences are evaporating in a new world consciousness: instantly communicating—an economically and politically interdependent planet.' The new generation, we are told, is ready to understand the ethos of reverence for natural systems intellectually, and to embrace it emotionally. The concluding chapter on metaphysics may be left to the philosophers.

STEPHEN PEPPER'S FIFTH WORLD HYPOTHESIS

Professor Laszlo is not alone in attempting a systems philosophy; Stephen Pepper, whose earlier work *World Hypotheses* is described in Chapter 1, has explicitly associated himself with the systems movement, both by his praise of Laszlo's work and by more recent work that incorporates the systems point of view.[38]

In 1967 Pepper published a large work *Concept and Quality,* adding a fifth world hypothesis to the four developed earlier.[39] This world hypothesis is based on the notion of a "purposive structure"—that is, the goal-seeking act.

This hypothesis is new in that it is not necessarily teleological. Nature need not be regarded as seeking a cosmic goal. All that is necessary is that this theory fully embody the "structural character of purposive activity."[40]

Pepper conceives of this world hypothesis as either a revision or extension of contextualism or as an entirely new hypothesis. The name suggested for this new hypothesis is *selectivism.* The author gives three initial reasons for selecting this root metaphor. First:

This act is the most highly organized type of simple purpose—possibly the most highly organized activity in the world of which we have any considerable evidence. It is the act associated with intelligence. And so it entails the features of the organism which performs the act. If we concentrate attention on this act, we are not likely to miss important features in cosmic structure and process. *For other activities and structures are likely to be simplifications of this.* [emphasis added]. We can learn about them by a sort of subtraction and will not be mystified by the demand for the addition of features emerging beyond our original basic categories.

The second reason for promise in our selection is that a purposive activity is one that may go on in the full illumination of consciousness. We can feel its whole qualitative course from initial impulse to terminal satisfaction. We can have the immediate feel of the perceptual demands of an environment in all its qualitative variety and graded intensity upon the search for the means of satisfaction. And we can feel the shock of a blocked anticipation when the wrong choice is made. We also feel here the emotions and have awareness of the values that seep into our perceptions in any living concrete act of pursuing a purpose within any environment that must be felt out, selected from, and cognized in purposive pursuit. By starting with a root metaphor that possesses full qualitative immediacy, this should go far towards resolving a problem that has tantalized mechanistic naturalism from Descartes down to the present—that of the relation of the mental to the physical—which cannot be dismissed as entirely a pseudo problem.

And third, this qualitative structure has submitted to a detailed conceptual analysis in behavioristic terms. Here, then, is an ideal opportunity to see how a set of effective and well elaborated concepts come to apply to a qualitative structure lived through in a man's immediate experience.[41]

Pepper chooses as a basic illustration a homely image: a person awakened by thirst, looking for a glass and water pitcher at the bedside, and not finding one, eventually getting up for a drink of water. The illustration shows all the essential features of an "appetitive purpose," although other appetitions may produce systems of greater complexity and range.[42]

An action of this type is lived and understood by everyone; its details are qualitatively accessible to all. The basis of the sequence of actions is a "drive," thirst. The drive evokes a reference to water; it becomes an "anticipatory set," or reference. "Drive references" are established—some cognitive, some not. A cognitive reference would be an image of water as quenching thirst. A noncognitive reference (or drive reference) would be the ultimate goal, termination of the drive through satisfaction. The cognitive references serve to designate subordinate goals or goal objects that are instrumental to the satisfaction of the drive. Turning on a light does not satisfy the thirst drive, but it is part of the chain of subordinate goals that will lead to satisfaction, as will other subordinate goals such as opening the cabinet and taking out a glass. The drive exhibits a "split dynamic," with references split into "anticipatory sets" leading through subordinate instrumental acts. The continuous wanting of the end goes on through and sustains the wanting of the means. In consequence of this split dynamics some instrumental acts prove to be erroneous or fruitless, and others prove to be correct. Thus the thirst drive sets up a chain of both physical and cognitive events, all organized for the purpose of satisfaction of the drive.

Pepper uses this image as the basis for an effort to coordinate both the qualitative element as experienced and the "conceptual" (i.e., behaviorist, psychological, and philosophical dimensions) elements of the process. Thus he generates two parallel lists of categories—the qualitative and the conceptual. The qualitative list contains concepts such as: *felt quality* (with dynamic urge for action); *duration* of the quality (yielding a continous qualitative strand); *intensity* of quality, goal reference, blockage, splitting of dynamic reference, selection of instrumental strands, and satisfaction. The conceptual categories, intended for an "objective" description, include parallel terms such as *bodily action* and *tension pattern, continuity through time, energy, vector character interaction with environmental qualities, vector changes, selection of response mechanisms, quiescence pattern.*

The stage is set for a vastly detailed analysis of acts and meanings set within the framework of a purposive act. The analysis is often on a high level of competence and accuracy of observation; but its main emphasis is on the epistemological dimensions of the concepts of purpose, goal, instrumental object, and of felt qualities, with the author laboring to maintain a place for the qualitative dimension within the scientific and epistemological view of the world. As such, the major portion of the book has no reference to social or sociological matters and, as such, is outside the scope of this study, except at a few important junctures. At one point Pepper argues that the entire process of scientific knowledge can be incorporated within this framework:

> . . . for a descriptive theory to work, provision has to be made for a dynamic system of references to connect the terms and relations of the description to the object described. And provision must also be made for those intermediate situations in descriptive knowledge where only partial but not full contact is made with the object described. We have this intermediate situation in distal perception. It is rather evident that we also have it in much scientific description . . . the final upshot of this analysis is that the ultimate object is a pattern of felt qualities, just as we found that the ultimate perceptual object must be.[43]

> In our view, then, a scientific hypothesis is an institutionalized set of references to an environmental object or type of object. The references are socially sanctioned and thereby obtain their dynamics as the operations performed by scientists in their field and laboratory work, operations which in large part have become traditional with investigators in the field.[44]

As a drive in an organism sets up a chain of anticipatory references, immediate objects, and ultimate objects, so the search for knowledge sets up a parallel chain of purposes including hypothesis, proximate scientific object, and ultimate scientific object. The affinity for systems theory is by now evident: Systems learn by acquiring information, gained in exchanges with the environment, and whether these systems are organisms, organizations, societies, or scientific disciplines matters little.

Again, Pepper's principal concern lies within the categories of contemporary science and philosophy, including epistemology, space, time, causality, and the rest, and so does not concern us here.

Pepper is aware of the problem of the cultural determination of language and, through language, of the influence of culture on the conferring of meaning upon percepts; his observations are accurate on a number of points, but the emphasis remains primarily epistemological with no cultural or historical materials offered.

But in his Chapter 15, "Values," Pepper finally reaches the problem of the social and cultural matrix of values. First, he anchors values in the purposive acts of organisms; positive values are generated by organisms as part of the chain of "appetitive" acts; negative values, similarly, as part of acts of aversion. In addition, affective and achievement values are likewise anchored in "purposive systems." Values involving pleasure, or the joy of achievement, are similarly related to the chains of purposive acts that encompass both intermediate and final goals. What, then, are the "selective systems" that generate social values?

There are three main groups of these—that of the social situation, that of the social institution, and that of the cultural pattern. The first is selective of *acts* performed by persons in a social situation. The third is selective social *dispositions* in the form of social institutions. Thus a cultural pattern is to a social situation as a personality structure is to a personal situation. The second, an institution within a cultural pattern corresponds to a role within a personality structure, and is likewise selective of a person's acts. Indeed, a role is precisely the embodiment of the demands of a social institution within a personality structure.[45]

Thus after long preparation, we arrive at a Parsonian systems scheme, complete with isomorphic parallelisms. As for the social situation as possible generator of values, Pepper tells us that moral values may appear to be generated within social situations extemporaneously, but the social situation is "just one selective system among many others," but not necessarily the decisive one even though virtually all human acts are performed "in view of a social situation. The decisive decisions, however, are often made by other selective systems and then contributed ready-made to the social situation over which they have a dominating influence. The fallacy then is to ascribe the selection solely to the agency of the social situation." Nevertheless, there may be many occasions in which the social situation "functions as a selective system with highest priority. Where the social situation is autonomous, the dynamics of the situation will be structured by the selection of systems of acts that will maximally satisfy the interests of the individuals involved. Thus the dynamics of the social situation is the result of the pressures involved and their effect on the configuration of tensions in the situation."

But more important as a selective system is a social institution. "A social institution is a dynamic dispositional structure that has its seat in the individuals who have been acculturated to it or are subject to such acculturation.

The seat will also include any artifacts needed for its functioning. Thus a military institution includes its arms and equipment. A religious institution includes its church or temple and ritual symbols."[46] Pepper reifies and personifies the social institution: "A social institution demands acts of conformity and selects acts for their conformity versus their nonconformity. It has three main kinds of sanction for enforcing conformity usually called (1) the official, (2) the approbative, and (3) the internal." The first sanctions are clear: the courts, police, priests and officials; the second are public approval and disapproval. The third are based in personality structures. They receive their dynamics from the habitual roles a person has adopted into his personality structure and from his conscience.

Here is where the selective system of personality structure comes to the support of institutional conformity. There is a close interplay between the two, since the origin of both roles and conscience as dynamic agents in the personality is found in social institutions by way of acculturation. For this is the process by which a child is taught the institutional structure of the society into which he is born. Institutions become dynamically embedded in the child's personality, so that almost or quite automatically he conforms to his institutions. The dynamics of his personality demands it. When acculturation is effective, there is no more need of the external sanctions of public approval or official compulsion. Actually a society maintains its solidarity largely through these internal sanctions. As a result of acculturation the members of a society just want to conform. It is only on exceptional occasions or in exceptionally rebellious persons that the external sanctions need to be applied.[47]

A social institution, then, becomes the basic model of a selective system. By maximizing satisfactions, the institutional values of dutifulness and conformity produce not mere conformity but an integrative harmony.

Furthermore, social integration is controlled by two separate and often antagonistic dynamics: the individualistic democratic versus the functional authoritarian, or the "open" versus the "closed" society. Pepper then summarizes these basic forces in a table of contrasting features, reminiscent of Mertonian-Parsonian concept-spaces:

The author does recognize that no society can be neatly classified along these dimensions. Some institutions in any one society may conform to the "functional-authoritarian' dimension, and others may conform to the "individualistic-democratic." He indicates further that where societies are subjected to greater external pressures, their institutions are more likely to conform to the first column; those under less pressure will conform to the qualities in the second column. Furthermore, he notes that cultural patterns may lack harmonious integration: Some institutions may show a cultural lag of one type or another, and conflicts among the institutions within a cultural pattern are likely to generate conflicts within personality structures of individuals.

Table 1:   Contrasts of Social Integrations

Functional Authoritarian Society	Individualistic Democratic Society
1. Survival as dominant motive	1. Happiness as dominant motive
2. Basic right of society over individual	2. Basic right of individuals and instrumental view of society
3. Centralization of government	3. Decentralization of government
4. Efficiency as chief aim of social organization	4. Opportunity for individual enterprise and satisfaction as aim of social organization
5. Discipline or team play as social attitudes sought	5. Initiative or tolerance as social attitudes sought
6. Duty or loyalty as personal attitudes sought	6. Satisfaction or compromise as personal attitudes sought

SOURCE: Stephen C. Pepper, *Concept and Quality–A World Hypothesis* (La Salle, Ill.; Open Court 1967), p. 541.

Though a disintegrated society may produce disintegrated personalities, a man may resist this influence; the correlation is not rigorous.

Thus more than 500 pages of close philosophical analysis seem to culminate in outmoded and simplistic social theory that no graduate student in sociology would expect to pass muster. Though we must be grateful that philosophers are at last beginning to turn their attention to the social and cultural dimensions of experience, no one can claim that they have even confronted—not to speak of having solved—the most elementary problems in social theory. The notion, for example, that societies are "held together" by the "demands" of institutions (where an institution appears now to be a "pattern" or "system" or a person with purposes) has been disputed. It has even been argued that society is "held together"—if that is the correct phrase—by the mutual exhaustion of conflicting parties or by the sense among conflicting groups that total conflict would be more damaging than partial compromises. But a state of temporary truces among some groups, partial struggles and conflicts among other groups, with truces being abandoned and resumed by various groups at any one time—none of this bespeaks an "integrative harmony" of values. One wonders what society Pepper is looking at when he says: "A society maintains its solidarily largely through these internal sanctions. As a result of acculturation the members of a society just want to conform. It is only on exceptional occasions or on exceptionally rebellious persons that external sanctions need to be applied." The notion

that persons in control of organizations or institutions may be as little social-
ized as those marginal to them has not occurred to him.

Nevertheless, Pepper is aware that the problem of values needs further
discussion, and he seeks to anchor values in one selective system of special
importance—natural selection. The lines of the argument are close to those
described by Laszlo: "Natural selection . . . generates values of its own." The
genes and chromosomes are the source of the governing dynamics of selection.
The split dynamics between ultimate and intermediate goals described earlier as part of all purposeful acts is seen here also. The split is between the
"overall governing energy" and the energy entering into the trial case (i.e.,
the individual mutation). Thus the values generated by this system are, positively, adaptation and, negatively, maladaptation or nonadaptation. With
man, of course, evolution took an abrupt turn: Natural selection has been
converted into cultural selection. It is cultural patterns of societies that have
greater or lesser fitness and adaptation. Thus the struggle for survival is a
struggle between cultural patterns. "Survival value now attaches to the acquired cultural pattern of the society rather than to the inherited structure of
the individual."[48] The dynamics of such patterns even determines the limits
of human obligations.

Pepper thus argues that natural selection makes still another contribution
to values: The limits of social obligation are determined by the boundaries of
the group organized to survive. The distinction made by anthropoligists between in-group and out-group exemplifies this development. Members of in-
groups, he tells us, are bound by culturally determined obligations. But members of the out-group may be regarded as less than human; obligations do not
apply to them.

The notion of universal humanity is recent; the threat of destruction posed
by atomic weapons requires us to extend the boundaries of obligation even
further to all men. But of course there are limitations; we need not extend
these boundaries to include animals, unless a new intelligent species evolves
that is able to communicate through symbols.

Thus, according to Pepper, natural selection supplies man with not only a
source of values, but also the determining factor that limits his obligations.[48]

This is not thought, but imagery of a confused kind. The first and most
obvious point is that values need not be anchored in naturalistic descriptions
of this kind; they may even be adopted in opposition to naturalism; there
may even be those who could choose the nonsurvival of mankind as an acceptable value, with no natural or selection process capable of proving they
are wrong. Since individual creatures all perish—and since, in the view of
evolution, all species are transitory and perish—could not evolution and natural selection be cited to "prove" that the purpose of natural systems is

death? Could not this purpose be cited as an ultimate value? There might even arise persons who refuse to extend the boundaries of their in-group to include those of out-groups, and might even do so in the face of atomic destruction. There is nothing but the equally "unscientific" value commitments of others to prove they are mistaken.

The image offered—man as having dominion over the earth and its creatures—is clearly one derived not from natural selection and adaptation but from still earlier sources. Since animals do not speak and have limited intelligence, no obligation exists in relation to them, with presumably minor exceptions for useful animals that can be domesticated. But persons oriented to what is today called ecological thinking can point out that the chain of life is a complex, sensitive, and interrelated totality, with man being highly dependent on mute plants and animals. Man has even been defined as a parasite on the cow, and there is much to be said for this definition; might not the cow and other such creatures be given a higher place than man in the scale of values? Might not there exist human obligations in this area? Pepper seems to take the concept of equal rights for granted, noting that members of out-groups are rarely regarded as legitimate candidates for such rights. But equal rights are surely not anchored in natural selection; the concept appears to be cultural, historical, limited in time and place, and so highly contingent.

But even here the ability to learn and to speak is not sufficient to determine the range of human obligation. Among human beings, do not some exist whose endowments are vastly greater than others? Would the "range of human obligation" extend equally to those who learn slowly and painfully compared to those of higher intelligence and adaptability? Can those of the first group be said to belong to the same species, having as little ability for survival as they show? Are they not a parasitic burden on the rest of "us?" Their right to survive is clearly arguable.

Thus the "argument' collapses at first touch in that the first logical development of Pepper's image lands the author in difficulties that cannot be resolved within the framework of his imagery. There is a concealed social point whose implications are not discussed: If survival is an ultimate value and human survival is based on cultural patterns, those better endowed with respect to learning and creating cultural patterns are clearly entitled to greater social claims than are those less endowed. But, of course, the concealed point of most systems theory can be described as an ideology having implications favoring intellectuals along the lines of what is often called a "meritocracy."

Perhaps these comments are premature, for Pepper now reaches the culmination of his discussion in a section called "Lines of Legislation for Human Values."

A selective system (and the values it generates) may under certain conditions exert control over another system (the "legislation" of one system over another). But these legislative relations of one system over others are not absolute but realtive, depending on circumstances. "Under certain conditions one selective system does become the focus of legislation, and this seems to confirm the correctness of the view of its ultimacy." In fact, important schools of ethics have been based upon virtually every selective system at one time or another, and its values given ultimate primacy. The next step is to list these selective systems and some of their associated values in a hierarchy based on their comprehensiveness:

Pepper tells us, with reference to this list: "The great empirical schools of ethics can almost be defined by reference to this list. (I pass over as unacceptable the nonempirical schools which rely upon appeals to a priori principles, or to indubitable 'goods' or 'value judgments' that are neither true nor false.)" The hedonists and utilitarians, the pragmatists and contextualists, the cultural relativists, the self-realizationalists ("from Plato to the Hegelians"), and "the evolutionary school of ethics," each place emphases on selected features of the chart; the first stress the affective, achievement, and pruden-

Table 2:   Lines of Legislation

Selected Values	Selective Systems	Positive and Negative Names for the Values
1. Affective	Consummatory and riddance fields	Pleasant vs. unpleasant (pain)
2. Conative-achievement	Appetitive and aversive purposive structures	Successful vs. unsuccessful (frustrated)
3. Prudential	Personal situation	Prudent vs. imprudent
4. Character	Personality structure	Personally integrated (responsible) vs. personally unintegrated (irresponsible)
5. Social	Social situations	Congenial or fitting vs. uncongenial or unfitting
6. Institutional	Social institution	Conformity vs. nonconformity
7. Cultural	Cultural pattern	Culturally integrated vs. culturally unintegrated
8. Survival	Natural selection	Adapted or adaptable vs. unadapted or unadaptable

SOURCE: Stephen C. Pepper, *Concept and Quality–A World Hypothesis* (La Salle, Ill.; Open Court, 1967), p. 552.

tial values; the second group (such as Dewey and Mead) stress the social situation; the third stress the cultural system, and so on. But there is no difficulty in bringing these conflicting positions into harmony. Thus, according to Pepper, only items 1, 2, and 8 possess "dynamics intrinsic to them." The other systems ("intermediate") are "activated in their choices by one or the other of these sources or by both." Depending upon the pressure of the environment, the impact of natural selection may be harsh or relaxed; man is cushioned from these pressures by the social structure, and the selective pressure of the environment is "felt by members of a society mainly as a social pressure." The greater the social pressure, the greater the impact of survival values will be upon a group, and the less social pressure, "the wider the range of satisfactions open to the purposive drives of the individual." Furthermore, if the social pressure increases, the more demand will occur for centralized integration and for social conformity to the institutions so integrated; as social pressure relaxes, the individual drives "acquire greater scope for their satisfactions in the gratifications of affection. . . . The prudent choice will now be more personal happiness and less stress for achievement. The personality structure becomes less duty bound."

Pepper offers no historical or cultural illustrations of these propositions; but any historian or sociologist will have no difficulty in finding illustrations that contradict them or that raise other problems. For example, the statement that the "other systems" contain no dynamics intrinsic to them would be firmly disputed. Pepper's contention is formally identical with the argument of Marxists concerning the superstructure and the infrastructure of society; Pepper dogmatically sweeps aside other views on this matter by not discussing them. There is no one-to-one correlation between cultural forms and institutions in societies under "high pressure" and those of societies under "low pressure." There might even be societies, such as American and Soviet societies, under which individuals are subjected to "high pressures" that do not emanate from threats to the survival of the society. In such a case the high pressures acting upon individuals might be justified as survival pressures without actually being such, a possibility not mentioned by Pepper. Is there a correlation between cultural forms in the arts, sciences, philosophy, religion, law, and so on and the level of survival "pressure" on a society? Many theorists would argue for their relative autonomy. Cultural forms appear to "evolve" along many lines and dimensions, with no traceable connection to the "survival pressure." The polarity seen by Pepper between natural selection on one side and purposive satisfaction on the other is resolved by the principle of social pressure. "The ideal social structure for a society is thus revealed as an adjustable one ready to centralize to whatever degree necessary to meet an emergency, and ready to decentralize for the maximum personal satisfaction of the individual consistent with security when an emergency is over." To be sure, this description is much too fluid; changes do not occur as smoothly as that; social traditions and personal habits "give a stability—not to say rigidity—to social institutions and personality structures. These demand consideration in their own right and resist sudden changes even though the environment would allow them." These appear to give these systems a dynamism of their own, which accounts for the "empirical theories" that focus their interest on these "intermediate" selective systems instead of on the true sources of the values. "Moreover, there is for some reason— perhaps simply because it is not a purely human power—a widespread resistance to recognizing survival values as values men are inevitably enmeshed in. So a moralist can focus on the social situations or on the cultural pattern as central for human values and indirectly recognize the impact of survival values by observing their effects as factors altering the social situation or constraining the structure of a society." The general solution, Pepper maintains, for these traditional difficulties lies in the recognition of the sequence of selective systems as described above, and of the "lines of legislation" running through them from the two opposite sources of "dynamic selectivity."

THE COSMIC RANGE OF VALUES

Values are anchored by Pepper in purposive systems, and values extend from "the purposive activities of intelligent organisms to social structures incorporating them." By including survival values, values extend thrrough the range of living forms, in terms of adaptation and adaptability. Are there values beyond these limits in the actual world? One can only speculate; mechanisms for maintaining steady states may exist in the inorganic world among molecular and atomic structures. If these prove to be self-corrective and comparable to the trial-and-error behavior of organisms and of genetic selection, "these inorganic structures would definitely be selective systems, and their selections value activities." Thus, if values prove to be generated down to the most primitive cosmic levels, this would be no great surprise. But relevant *human* values appear to be limited to the selective systems described above.

Systems philosophers such as Pepper and Laszlo clearly seek to anchor ethical values in empirical matters. In the process they sweep aside the still unresolved problems in this area. Although the following material is or should be elementary, the systems philosophers appear never to have confronted it. The first is the difficulty involved in basing ethical "systems" on empirical matters. The California philosopher W. T. Jones states the problem:

Much evidence has been accumulated that ethical and political beliefs are relative to social class, economic status, and, even, to professional group. It would seem, then, that the Founding Fathers, so far from formulating universally and eternally true moral axioms, merely gave expression to the prejudices of their own small social class in an English-speaking culture. It is true that in the welter of diverse customs and codes uniformities can be found—"universals" some anthropologists call them. But these universals are not at all the ethical absolutes of the Founding Fathers. They are empirical generalizations describing the sorts of things men value, not assertions of the objective validity of the values so affirmed. Such propositions assert not an "ought," but an "is"—even if this happens, in some cases to an is-about-an-ought.

We have indeed been driven to recognize that there is a basic distinction between saying (a) "all men ought to respect the rights of others," and (b) "in all societies we encounter the belief that all men ought to respect the rights of others." The latter proposition, even if true (which seems unlikely), would not establish the former. For the former is an ethical assertion; the latter is a factual one. We know the kind of evidence that would tend to verify or diversify the latter; it consists in empirical observations . . . But what sort of evidence would be relevant to the former (ethical) assertion? . . .

But further: the universals (or better, the uniformities) formulated in the current, empirical approach to values are, characteristically, relations between the values that people experience or affirm, and either (a) certain biological drives . . . or (b) the facts of social interaction. . . .

These are typical empirical generalizations—"is" statements, not "ought" statements. Such universals (or uniformities), far from establishing favored ethical absolutes, tend rather to undermine the very conception of there being any ethical absolutes at all.[51]

The discussion above is far from being the first or only statement of the problem. Sociologists (but apparently not philosophers) are familiar with Max Weber's statements on the subject, made almost 60 years ago.

. . . I am most emphatically opposed to the view that a realistic "science of ethics," i.e., the analysis of the influence which the ethical evaluations of a group of people have on their other conditions of life and of the influences which the latter, in their turn, exert on the former, can produce an "ethics" which will be able to say anything about what *should* happen. . . .

The empirical-psychological and historical analysis of certain evaluations with respect to the individual social conditions of their emergence and continued existence can never, under any circumstances, lead to anything other than an *"understanding"* explanation. . . .

In almost every important attitude of real human beings, the value-spheres cross and interpenetrate. The shallowness of our routinized daily existence consists indeed in the fact that the persons caught up in it do not become aware, and above all do not wish to become aware, of this partly psychologically, part pragmatically conditioned motley of irreconcilably antagonistic values. They avoid the choice between "God' and the "Devil" and their own ultimate decision as to which of the conflicting values will be dominated by the one, and which by the other.

The fruit of the tree of knowledge, which is distasteful to the complacent but which is, nevertheless, inescapable, consists in the insight that every single important activity and ultimately life as a whole, if it is not to be permitted to run on as an event in nature but is instead to be consciously guided, is a series of ultimate decisions through which the soul . . . chooses its own fate, i.e., the meaning of its activity and existence. Probably the crudest misunderstanding which the representatives of this point of view constantly encounter is to be found in the claim that this standpoint is "relativistic"— that it is a philosophy of life which is based on a view of the interrelations of the value-spheres which is diametrically opposite to the one it actually holds, and which can be held with consistency only if it is based on a very special type of ("organic") metaphysics.[52]

These brief extracts do not do justice to Weber's carefully knit argument, but at least they call attention to an issue that contemporary philosophers, systems philosophers among them, have evaded. Certainly the claim of systems philosophers to have resolved problems of ethics and values, by means of systems thinking, cannot be taken seriously and even raises questions, if other considerations had not also done so, of their competence as philosophers. Weber appears to have anticipated and dismissed more than 50 years ago the possibility of an "organic metaphysics," which the systems philosophers bring forth and which in fact is something quite old.

By now the reader may sense rightly, that the philosophical competence of systems theory is not of a high order. And by now the vices of this way of thinking are clear: Systems thinkers exhibit a fascination for definitions, conceptualizations, and programmatic statements of a vaguely benevolent, vaguely moralizing nature, without concrete or specific references to historical, social, or even scientific substance. They behave as camp followers of "science," not of the sciences, picking up whatever details serve to illustrate their views, while disregarding others. They collect analogies between the phenomena of one field and those of another (preferring to call them isomorphisms, though the difference is not discernible), the description of which seems to offer them an esthetic delight that is its own justification.

But the most telling objection, at least as far as we have gone, is that systems theory achieves its all-encompassing "universality" only by its very abstractness: All things are systems by virtue of ignoring the specific, the concrete, the substantive; No evidence that systems theory has been used to achieve the solution of any substantive problem in any field whatsoever has appeared. The peculiar vagueness and eclecticism of systems thinking is beautifully caught in the image offered by Ida R. Hoos:

. . . the approach, as we now encounter it, resembles the geological phenomenon known as "Roxbury pudding-stone" in both history and constitution. This formation, located in a suburb of Boston, Massachusetts, resulted from glacial movement, which over the miles and centuries dragged with it, accumulated, and then incorporated a vast heterogency of types of rock, all set in a matrix and solidified in an agglomerate mass. Many fragments still retain their original identity and character; some have undergone metamorphosis in varying degrees. In like manner, the systems approach is a kind of mosaic, made up of bits and pieces of ideas, theories, and methodology from a number of disciplines, discernible among which are—in addition to engineering— sociology, biology, philosophy, psychology, and economics.[55]

Though her description was written primarily with respect to systems thinking as operations research, we find that it applies as well to systems thinking in its loftier guise as philosophy.

Another objection applies to systems philosophy as philosophy, rather than as social doctrine. Systems philosophy remains philosophically passive and subservient to the sciences, not as sciences but as sources of philosophical orientation. This is not a trivial objection, for the sciences themselves prove to labor under philosophical difficulties that remain unresolved, and if systems philosophy looks to science for answers, the problems are likely to remain unresolved. Systems theorists will continue to announce new beginnings, new orientations, new approaches; will continue to discuss the historical forerunners of systems theory, the epistemological problems of philosophy; and will continue to sharpen their conceptual and methodological tools without applying them. Thus they fail to do substantive work in specific scientific areas and fail to do genuine philosophic work that is waiting to be done and that is not dependent on scientific "data." Speculation as speculation can be valuable and provocative, just as empirical data are valuable, but the hybrid mixture of fact and speculation which is systems theory weakens both.[54]

As theory, systems philosophy is a mixture of speculation and empirical data that is satisfactory as neither. The speculative materials merely reinterate the categories of the Cartesian world view without going beyond them, and the empirical data are a hodgepodge drawn on no consistent principles from any and every field.

As practical technique, in its various applied forms as operations research, systems analysis, and the like, we see that it does not work in the social or political arena. Where it has failed in application, we are told that the theory is fine but the applications were poorly carried out. This of course draws the rebuke that if a technique is not judged by its applications, how then can it be judged? Furthermore, how a theory that is not borne out by its applications still retains significance as a theory remains unexplained.

A tentative defense is that the systems analysts and philosophers are more at home within their specialized fields and that their excursions into the social and cultural worlds were ill-advised; systems theory might be essentially sound, but needs the knowledge of the social scientist (and there are social scientists of the systems persuasion).

In this chapter we have examined the emergence of systems theory in a number of disciplines and have seen the practitioners of those disciplines extend their claims to the social, cultural, political and economic. Chapter 7 is an examination of the "migration" of systems thinking into the social world—political science, sociology, psychiatry, social work, and urban planning. Social scientists with detailed substantive knowledge of their fields have adopted at least the vocabulary of systems thinking. Perhaps, they will show its fruitful applications.

NOTES

1. Ludwig von Bertalanffy, *General System Theory—Foundations Development Applications* (New York: Braziller, 1968), pp. vii–viii.
2. Laszlo's works include: *Introduction to Systems Philosophy—Toward a New Paradigm of Contemporary Thought,* with a Foreword by Ludwig von Bertalanffy (New York: Jordon and Breach, 1972); *The Systems View of the World* (New York: Braziller, 1972); in addition, he is editor of the *International Library of Systems Theory and Philosophy* (published by Braziller), whose titles include: Ervin Laszlo, ed., *The Relevance of General Systems Theory—Papers Presented to Ludwig von Bertalanffy on His Seventieth Birthday* (New York: Braziller, 1972); H. H. Pattee ed., *Hierarchy Theory—The Challenge of Complex Systems;* Ervin Laszlo, Ed., *The World System—Models, Norms, Applications;* John W. Sutherland, *A General Systems Philosophy for the Social and Behavioral Sciences.* His works bear high praise from Von Bertalanffy, the physicist Henry Margenau, and the philosopher Stephen Pepper, among others; his writings, then, may be taken as representative of systems thinking as philosophy.
3. Laszlo, *Introduction to Systems Philosophy,* p. ix.
4. *Ibid.,* p. 3.
5. *Ibid.,* p. 4.
6. *Ibid.,* pp. 5–6.
7. *Ibid.,* p. 6.
8. *Ibid.,* pp. 6–10.
9. *Ibid.,* p. 11.
10. *Ibid.,* p. 12.
11. For a brief description of Pepper's ideas Chapter 1, this volume. Pepper's newest world hypothesis is described later in this chapter.
12. Laszlo, *Introduction to Systems Philosophy,* pp. 15–16.
13. *Ibid.,* p. 19.

14. *Ibid.*, p. 16.
15. Laszlo, *The Systems View of the World*, pp. 67–68.
16. Laszlo, *Introduction to Systems Philosophy*, Chapter 3, especially p. 29.
17. *Ibid.*, p. 32.
18. *Ibid.*, pp. 38–41.
19. *Ibid.*, p. 42.
20. pp. 35–36; the scheme is repeated for all physical, biological, and social levels; see pp. 37–98.
21. *Ibid.*, pp. 98ff.
22. *Ibid.*, pp. 99–100.
23. *Ibid.*, p. 95.
24. *Ibid.*, pp. 112–113.
25. Max Weber, "Bureaucracy," in H. Gerth and C. W. Mills, Eds., *From Max Weber* (New York: Oxford University Press, 1958), pp. 228–229.
26. Quoted in L. Coser and B. Rosenberg, Eds., *Sociological Theory*, 2nd ed. (New York: Macmillan, 1964), p. 473. The original source is given as Max Weber, *Gesammelte Aufsätze zur Soziologie und Sozialpolitik*, pp. 412ff.
27. Laszlo, *Intro. to Systems Philosophy*, p. 114.
28. *Ibid.*, pp. 116–117.
29. *Ibid.*, p. 99.
30. *Ibid.*, p. 249.
31. *Ibid.*, pp. 249–250.
32. *Ibid.*, p. 254. By the phrase "social differentiation is not anathemic but propaeduetic to individual freedom," the author appears to mean that social differentiation is not inimical but conducive to human freedom.
33. Max Weber, "Science as a Vocation," in H. Gerth and C. W. Mills, Ed., *From Max Weber* (New York: Oxford University Press, 1958), pp. 148–149.
34. Laszlo, *Introduction to Systems Philosophy*, pp. 258–259.
35. *Ibid.*, p. 264.
36. *Ibid.*, pp. 266–268.
37. *Ibid.*, pp. 270–271.
38. The dust jacket of Laszlo's *Introduction to Systems Philosophy* bears the following comments by Pepper. "Laszlo's work comes as a breath of fresh air, like opening a window in a crowded, smoke-filled room. It is a world hypothesis thoroughly empirical without any dependence on items of self-evidence or indubitability. Its paradigm or root metaphor is system or, more specifically, the dynamic self-regulating system. This is a happy choice, and possibly the most fruitful or even the correct one for a detailed synthetic comprehension of the structure of the universe. It seems applicable to the full range of empirical material available through the sciences, the arts and elsewhere, and it can utilize the results and methods of the extensive development of systems theoretical analysis." The work is also praised by Henry Margenau.
39. Stephen C. Pepper: *Concept and Quality—A World Hypothesis* (La Salle, Ill.: Open Court, 1967).
40. *Ibid.*, p. 15.
41. *Ibid.*, pp. 17–18.
42. *Ibid.*, pp. 19–20.
43. *Ibid.*, pp. 291–292 and p. 306.
44. *Ibid.*, p. 288.
45. *Ibid.*, pp. 534–535.
46. *Ibid.*, p. 537.
47. *Ibid.*, pp. 538–539.
48. *Ibid.*, pp. 544ff.

49. *Ibid.,* pp. 548–551.
50. *Ibid.,* p. 553.
51. W. T. Jones, *The Sciences and the Humanities—Conflict and Reconciliation* (Berkeley and Los Angeles: University of California Press, 1967), pp. 15–16. In the preface to this work the author expresses his thanks for many vigorous discussions of the manuscript, including discussion with Stephen Pepper. Professor Pepper might, then, have explicitly confronted or at least discussed the difficulties inherent in his position which represents a form of the dogmatism he deplores in others.
52. Max Weber, "The Meaning of Ethical Neutrality" in *The Methodology of the Social Sciences,* trans. by Edward A. Shils and Henry A. Finch (New York: Free Press, 1949), pp. 13–14, 15–16, and 18.
53. Ida R. Hoos, *op. cit.,* p. 27.
54. W. T. Jones, *op. cit.,* p. 31: "While philosophy is to some extent concerned with finding out what is real, it is also, and chiefly, an attempt to find an appropriate language for describing what everybody already agrees to be real."

Systems Thinking in the Social Sciences

$\mathbf{T}$he influence of systems thinking on sociology, political science, and anthropology has weakened the lines of demarcation between these disciplines in specific areas of contact. The influence of Talcott Parsons on political science has been at least as great as his influence upon sociology. Similarly, structural-functionalism as a theoretical movement has moved anthropology[1] and sociology closer together, and political scientists and anthropologists have converged in using an evolutionary framework for examining political forms in "primitive" societies. Finally, the newer systems-cybernetic movement has produced responses in all three fields. As a result these overlapping and mutual influences have worked to make the distinctions between academically separate fields somewhat arbitrary, especially from the systems point of view. The separation adopted in this chapter serves only to impose some order on the materials.

SOCIOLOGY

THE SOCIAL SYSTEM OF TALCOTT PARSONS

In one sense the "social system" of Talcott Parsons should not be included under the general heading of systems theory because Parsons developed his conception of the social system at least two decades before systems thinking as we have described it emerged as a significant movement. Parsonian sociology dates from the 1930s; cybernetics and the rest emerged in the 1950s and can be roughly correlated with the rise of the computer. In addition, advocates of the later systems movement claim that their approach differs significantly from Parsons' in several important dimensions. But in another sense there are definite affinities between the two:

1. Both Parsonians and system-cybernetic theorists claim a common ancestry in the work of Henderson, Pareto, Cannon, and others.

2. There are internal resemblances between cybernetic-system theory and the Parsonian social system.

3. Parsons has explicitly embraced modern systems theory, or at least important tenets of it, in his recent writings; in the Preface to his *Societies— Evolutionary and Comparative Perspectives,* he indicates that the comparative approach to the study of societies "posed the problems of evolution with renewed definiteness and urgency. This has also been stimulated by new developments in the unification of scientific theory, particularly in this case between the biological and the social sciences."[2] Parsons treats social evolution as analogous to organic evolution:

Symbolically organized cultural patterns, like all other components of living systems, have certainly emerged through evolution . . . the more general cultural patterns provide action systems with a highly stable structural anchorage quite analogous to that provided by the genetic materials of the species-type, focusing on the learned elements of action just as the genes focus upon the inheritable elements.[3]

The equation is still further developed:

A fundamental principle about the organization of living systems is that their structures are differentiated in regard to the various exigencies imposed upon them by their environments. Thus the biological functions of respiration, nutrition-elimination, locomotion, and information-processing are bases of differentiated organ-systems, each of which is specialized about the exigencies of certain relations between the organism and its environment. We will use this principle to organize our analysis of social systems. . . .

We have referred to a hierarchy of control which organizes the interrelations of the analytically distinguished systems. This includes the *cybernetic* aspect of control by which systems high in information but low in energy regulate other systems higher in energy but lower in information. . . . Thus, a programmed sequence of mechanical operations (e.g., in a washing machine) can be controlled by a timing switch using very little energy compared with the energy actually operating the machine's moving parts or heating its water. Another example is the gene and its control over protein synthesis and other aspects of cell metabolism.

The cultural system structures commitments vis-a-vis ultimate reality into meaningful orientations toward the rest of the environment and the system of action, the physical world, organisms, personalities, and social systems. In the cybernetic sense, it is highest within the action system, the social system ranking next, and personality and organism falling respectively below that. . . . Thus we must focus on the cybernetically higher-order structures—the cultural system among the environments of the society—in order to examine the major sources of large-scale change.[4]

In Parson's companion volume, *The System of Modern Societies,* the relation to systems thinking is maintained:

The present book is written in the spirit of Weber's work but attempts to incorporate the developments in sociological theory and other fields of the past fifty years. One important difference in perspective has been dictated by the link between organic evolution and that of human society and culture. Developments in biological theory and in the social sciences . . . have created firm grounds for accepting the fundamental continuity of society and culture as part of a more general theory of the evolution of living systems.[5]

Parsons conceives of social systems as "constituents of the more general system of action, the other primary constituents being cultural systems, personality systems, and behavioral organisms. . . . The distinctions among the four subsystems of action are functional. We draw them in terms of the four primary functions which we impute to all systems of action, namely pattern-maintenance, integration, goal-attainment, and adaptation." The social system has primacy in the integrative function; primacy of pattern-maintenance is attributed to the cultural system; goal-attainment applies to the personality of the individual, and adaptation applies to the behavioral organism.

It is at this point that Parsons explicitly allies himself with systems thinking, if his earlier references to cybernetics had not already made the point:

In analyzing the interrelations among the four subsystems of action . . . and between these systems and the environments of action . . . it is essential to keep in mind the phenomenon of *interpenetration.* Perhaps the best known case of interpenetration is the *internalization* of social objects and cultural norms into the personality of the individual. Learned content of experience, organized and stored in the memory apparatus of the organism, is another example, as is the *institutionalization* of normative components of cultural systems as constitutive structures of social systems. We hold that the boundary between any pair of action systems involves a 'zone' of structured components or patterns which must be treated theoretically as *common* to *both* systems, not simply allocated to one system or the other. For example, it is untenable to say that norms of conduct derived from social experience, which both Freud (in the concept of the Superego) and Durkheim (in the concept of collective representations) treated as parts of the personality of the individual, must be *either* that *or* part of the social system . . .

It is by virtue of the zones of interpenetration that processes of interchange among systems can take place. This is especially true at the levels of symbolic meaning and generalized motivation. In order to 'communicate' symbolically, individuals must have culturally organized common codes, such as those of language, which are also integrated into systems of their social interaction. In order to make information stored in the central nervous system utilizable for the personality, the behavioral organism must have mobilization and retrieval mechanisms which, through interpenetration, subserve motives organized at the personality level.

Thus, we conceived social systems to be 'open,' engaged in continual interchange of inputs and outputs with their environments. Moreover, we conceive them to be internally differentiated into various orders of subcomponents which are also continually involved in processes of interchange.[6]

It appears, then, that despite the more or less independent emergence of systems thinking in other disciplines, Parsons' "system" may be classified as belonging to systems thinking; certainly its philosophical categories and its metaphysics are the same.[7]

The "highly stable structural anchorage" provided to "action systems" by cultural patterns, which Parsons sees as "quite analogous to that provided by the genetic materials of the species type," may be viewed with some skepticism at this point in history.

WALTER BUCKLEY

Among sociologists, Walter Buckley is perhaps the most outspoken representative of the General Systems Theory Movement. His *Sociology and Modern Systems Theory* is explicitly intended to replace all earlier and outmoded conceptions of society with the systems view, and his anthology of readings *Modern Systems Research for the Behavioral Scientist* is an authoritative and ambitious collection of canonical writings, having almost "official" status in the systems movement.[8]

His purpose in *Sociology and Modern Systems Theory* is not only to explain and stimulate interest in systems concepts such as feedback, input, output, boundary, systems, and the like, but also to show their fruitfulness for sociological theory. But before this can be done, some rubbish has to be cleared away, rubbish in the form of outmoded sociological concepts. The major portion of the book is devoted to a survey of contemporary sociological theory, designed to show that where that theory has not been completely immobilized and sterilized by the dead hand of the past, it is moving in the direction of systems thinking. Thus the subject matter of the book is not society seen from a systems point of view, but rather sociological theory seen from the systems point of view.

Buckley begins by noting that sociological theory has for some time been living off the intellectual capital of previous centuries. Consequently, the dominant mode of thinking has been along the lines of "equilibrium, consensus, or functional theory" and has also given rise to the current (mid-1960s) rise of criticism of that theory on the grounds of its being either mechanical or organismic in nature, models that are inappropriate for the study of culture and society. But, he indicates, since World War II there has been intellectual ferment in the sciences that is finally having influence upon sociological thinking—"cybernetics, information and communication theory, general systems research, and the like." Thus attention is shifting from "eternal substance" and energy transformation to a focus on organization and its dynamics "based on the 'triggering' effects of information transmission. Here is to be found the secret of distinguishing living from nonliving matter, adaptive, morphogenic processes from equilibrating, entropic processes." The "interdisciplinary generalizing and integrating potential of this newer systems theory has been widely accepted and utilized in all the major scientific fields, but sociology has remained untouched by it." [9]

The aim of the book is to "investigate the principles and methods of modern systems research as the basis for a more adequate model or theoretical framework of the socioculture system."

Buckley then moves to a brief description of systems concepts that are by now familiar: the relation of wholes and parts, the anti-entropic nature of the "higher" systems—biological, psychological, and sociocultural. These latter are distinguished by their "morphogenic" properties. Rather than minimize organization or maintain a given structure, "they typically create, elaborate, or change structure as a prerequisite to remaining viable, as ongoing systems." The problem is to understand what gives them this capability. Thus we arrive at the cybernetic principles of control, positive and negative feedbacks, communication and information processing, and the like, among them goal-seeking, self-awareness, and self-direction.

Sociocultural researchers, Buckley tells us, must learn to develop and apply these concepts as analytic tools, in place of the older and cruder techniques of correlation and factor analysis.

Such concepts will enable us to see more clearly "the inadequacies of the current 'consensus' model and its relatively static conception of 'institutions,' 'social control,' and social 'order' and 'disorder.'"

Cybernetic models of society can arrange these matters better than can the earlier "equilibrial" or "homeostatic" model, and in fact, Buckley indicates, some of the better conceptualizations going on in current sociology (e.g., on deviance generation) are excellent examples of the modern systems approach whether consciously recognized or not.

CRITIQUE OF EARLIER SOCIAL SYSTEM MODELS

Before presenting modern system theory for sociology, Buckley attempts to clear away the underbrush by a critical examination of earlier models of social systems, among them the mechanical and the organic.

The mechanical model of society, prevalent in the eighteenth and first half of the nineteenth centuries, was based on the rationalist physics of the seventeenth and eighteenth centuries. Concepts of social space, social coordinates, attraction, inertia, centrifugal and centripetal social forces, pressures, social atoms, and the like all reveal their provenance. Social thinkers in the late nineteenth century added concepts such as fields of force, transformation of energy, and social entropy. One important exception was Pareto, whose rational mechanics was based not on specious analogies but on more general mechanical principles that seemed to apply to social phenomena.

Thus, we have at base the concept of "system," of elements in mutual interrelations, which may be in a state of "equilibrium," such that any moderate changes in the elements or their interrelations away from the equilibrium position are counterbalanced by changes tending to restore it.

It is this conception that has been taken over almost unchanged by many contemporary sociologists, notably George C. Homans, and Talcott Parsons (both influenced by Henderson at Harvard). . . .

Many social scientists referred to society or the group as a "social system" without, however, taking a wholly mechanistic view—in fact, like MacIver, placing an opposite emphasis on the mentalistic factors.[10]

Buckley takes exception to Parsons' value and unclarified use of this model, especially in his use of the concepts of equilibrium and inertia. "Furthermore, when Parsons goes on to admit change forces *endogenous* to the system, we part company with anything recognizable to the student of classical mechanics. As others have noted, to say that internal system forces tend toward equilibrium but in fact may lead to change is a contradiction in terms."

In turning to the organic model, Buckley finds this equally faulty. The organic metaphor for society is quite old, but serious employment of the model in scientific terms is traceable to Herbert Spencer and his Social Darwinist followers. According to Buckley, Lester Frank Ward demolished Social Darwinism by his insightful emphasis on process: knowledge-attaining processes, struggle for structure, difference of potential, and synergy (the organic working together of antithetical forces in nature). These insights, Buckley indicates, are in the mood of modern systems theory.

The modern version of the biological model of society is represented by functionalism, especially by those who today emphasize order, cooperation, and consensus.

Thus Parsons, in his functional analysis of social change, after representing the social system as tending to maintain a relatively stable equilibrium by way of continuous processes which "neutralize" endogenous and exogenous sources of variability that would change the structure if proceeding too far, then gives an *organismic* illustration of *homeostasis:* temperature regulation in animals. . . . The basic point here is that whereas mature organisms, by the very nature of their organization, cannot change their given structure beyond very narrow limits and still remain viable, this capacity is precisely what distinguishes sociocultural systems.[11]

On these and other lines, Buckley rejects Parsons' social system. It is a vaguely conceptualized amalgam of the mechanical (equilibrium) and the organismic (homeostasis) models, placing excessive emphasis on normality and stability, and devaluing change, conflict, strife. Parsons' work is critized as methodologically faulty. "Change" may be such only from the point of view afforded by a reference frame that selects only one aspect of the system the other homeostatic mechanisms are tending to maintain. This has led Parsons into an illusory normative point of view that he confuses with social processes. Buckley, then, appears to reject Parsons' recent reapproachment with systems-cybernetic theory; Parsons' work is based on earlier, and faulty models, despite Parsons' claim that he was really saying the same thing all along.

From our point of view, both the differences and the affinities are evident: Parsons' work does predate modern systems theory, and the latter does contain terminological innovations not to be found in Parsons. However, the new terminology does not alter the underlying sameness of their concepts; rather they appear to be improvements and refinements of what is implicit in Parsons' work; the dispute may be merely a sectarian quarrel over primacy. The underlying sameness consists in that something called "system" is reified and endowed with cosmic meanings and purposes. Further, although Parsons himself may have overstressed stability and pattern-maintenance, there is no reason why his system may not accommodate the concept of change. In his *Societies—Evolutionary and Comparative Persepectives* Parsons himself remarks:

The special type of process with which this book is concerned, however, is *change*. Though all processes change something, it is useful for our purposes to distinguish from others the processes which change social structures. Here, it is evident that many complex processes are necessary to *maintain* the functioning of any societal system; if its members never did anything, a society would very soon cease to exist.

At the most general theoretical levels, there is no difference between processes which serve to maintain a system and those which serve to change it. The difference lies in the intensity, distribution, and organization of the "elementary" components of particular processes relative to the states of the structures they affect.[12]

While recognizing the terminological differences between Parsonian system theory and modern systems-cybernetic theory, I suggest that they can be grouped together, especially in view of Parsons' adoption of the cybernetic point of view; at the same time Buckley's critique does score significant though not original points.

Buckley then criticizes modern functionalism, especially in anthropological theory, as a misuse of the organismic model, suitable for only superficial analysis of "sociocultural" systems.

Before moving to a detailed description of Parsons' and Homans' work, Buckley mentions favorably a third model, the Process Model, which is described as having been influential in early twentieth-century American sociology. The Chicago school is portrayed as especially influential, including Albion Small, George Herbert Mead, Robert Park, and Ernest W. Burgess. Their "perspective," Buckley indicates, was not systematically developed to the point of being called a model and even at times included both the organismic and the equilibrium models, but they had the great merit of seeing through their weaknesses; in fact, the process view, according to Buckley, is "very congenial to—even anticipative of—basic principles of cybernetics."

In essence, the process model typically views society as a complex, multifaceted, fluid interplay of widely varying degrees and intensities of association and dissociation. The "structure" is an abstract construct, not something distinct from the ongoing interactive process but rather a temporary, accommodative representation of it at any one time. These considerations lead to the fundamental insight that sociocultural systems are inherently structure-elaborating and changing; for some, the terms "process" and "change" were synonymous.[13]

Other process theorists worthy of note include Marx, Engels, Albion Small, Cooley, Simmel, Leopold von Wiese and Howard Becker, W. I. Thomas and Florian Znaniecki, though Buckley notes that among many of these the insights and formulations were often groping, partial, or tentative. Buckley makes evident his sympathies with the process model and its diverse representatives and claims them as forerunners of general systems thinking in the social sciences; a very different lineage is claimed than the line from Pareto to Henderson, Cannon, and Parsons.

THE PARSONS AND HOMANS MODELS

Buckley's critique of the Parsonian model—especially as embodied in *The Social System*[17] and in Parsons' and Shils' *Toward a General Theory of Action*—follows familiar lines: Parsons overemphasizes order and control and tends to label change as problematic—that is, as deviance or in some way threatening. He evokes "control mechanisms," institutionalized value systems, and he does so in a normative way that disregards the existence of the pathological, the conflictual, the contradictory. Thus Parsons betrays a conserving orientation. His system is selective, anthropomorphic, ridden with teleology, neglects systemic stresses and strains, cannot explain change, and is an uncomfortable mixture of the equilibrium and homeostatic models.

Buckley appears to regard Homans' work as being of much higher caliber. He notes that although Homans starts from the same equilibrium concept as did Parsons, he constructs principles and conclusions that are quite different.

Like Parsons, Homans sees a system as defined in terms of "determinate, reciprocal relationships of *all* its parts, regardless of the structure in which these interrelations are manifested." (The reader will recognize the resemblance to Ashby's views on cybernetic functions.) These elements may be activities, interactions, sentiments, or norms. These interrelations may be seen as a family structure, a work structure, or a community structure. "There is no attempt to take any of these structures as fixed, privileged points of reference." Therefore, deviance, strain, conflict are as integral to the system as are harmony, cooperation, and order. The maintenance of a pattern is a problem or a miracle. A social system is a configuration of dynamic forces, sometimes in balance (steady state), sometimes out of balance, such that change occurs. The system does not *impose* control, it is the control. Systems do not seek equilibrium and do not have problems; they do not give rise to structures because they are "needed" by the system. Homans rejects the structure-function model based in biology.

But though Homans' model avoids the difficulties and ambiguities of Parsons', it has difficulties of its own: its weakness is in a "mechanically derived notion of equilibrium." Homans gives us no basis for judging whether a system is or is not in equilibrium; further, Homans even approaches modern systems theory in arguing that mechanical systems do not elaborate structure, do not reach new levels of survival, and do not "manifest efficient causes playing into the hands of final causes." Buckley concludes his description with an expression of disappointment:

It is clear that Homans has left both the classical equilibrium model and the organismic functional model far behind in the evolution of his theoretical conception of the socioculture system. We shall have occasion to discuss later his disappointing appeal to a reductionist psychologism—a giant-step down from his earlier systems approach.[14]

THE GENERAL SYSTEMS PERSPECTIVE

Buckley presents general systems theory as the "next" step in sociological thinking and as the logical heir of previous developments. He suggests six points that should make the modern systems approach especially attractive to sociology; this approach promises to develop:

1. A common vocabulary unifying the several "behavioral" disciplines.
2. A technique for treating large complex organizations.
3. A synthetic approach where piecemeal analysis is not possible due to the intricate interrelationships of parts that cannot be treated out of context of the whole.
4. A viewpoint that gets at the heart of sociology because it sees the socio-cultural system in terms of information and communication nets.
5. The study of *relations* rather than "entities," with an emphasis on process and transition probabilities as the basis of a flexible structure with many degrees of freedom.
6. An operationally definable, objective nonanthropomorphic study of purposiveness goal-seeking system behavior, symbolic cognitive processes, consciousness and self-awareness, and sociocultural emergence and dynamics in general.[15]

These are of course large promises; they indicate where work is to be done, rather than achievements. Thus, to the mechanical equilibrium model and to the organismic homeostatic model, Buckley juxtaposes a third model, the process, or adaptive system, model.

INTERACTIONS, INSTITUTIONS, AND DEVIANCE

Buckley's third chapter on systems needs no summarization; it presents the basic tenets of cybernetic-system theory as we have seen it unfold in its sources, with primary emphasis on von Bertalanffy's general system theory, information and communication theory, and cybernetics as developed by Wiener and Ashby, combined with references to social science thinkers who have adopted this point of view. In the last three chapters of his book, Buckley attempts to recast major trends in sociological theory within a system framework, by showing that contemporary emphases on dynamic processes are in effect application or anticipation of systems thinking.

ACTS AND INTERPRETATIONS

Buckley's system theory proceeds on two levels: a "microsociological" level, in which the individual personality emerges as a system of potential actions and social meanings emerge out of the interactions of individuals in face-to-face interchanges. Buckley adopts information theory bodily as a theoretical model here; organization, whether of personality or of social interaction systems, proceeds by the selection—from a universe of possible actions or meanings—of specific actions or meanings with increasing probabilities, where the subset chosen is meaningfully related to the environmental constraints confronting the individual or the group. Thus, Buckley places social forms and their origin (morphogenesis) within the matrix of actions and interactions.[16]

Buckley credits John Dewey and especially George Herbert Mead with originating a "process view" of social interaction that was basically a system approach. A social act is explainable not merely in terms of stimulus-response but as an ongoing complex of acts and responses; a stimulus does not "produce" a response; the context leads the individual to select both stimuli and responses and generates responses within himself as well as within those with whom he interacts. The system of responses evoked by the individual within himself becomes the organized "self."

Furthermore, the interactional field itself is seen as a complex set of interchanges out of which meanings and personalities emerge, especially the simplest dyadic fields. Sociology itself is returning to this view in the various forms of exchange theory now emerging, among them Homans, Parsons, Blau, Coleman, and Newcomb.

Whichever models are examined, the primary concept is that of an interactional field: an interpersonal matrix involving persons, objects, values, and exchanges.

Thus socialization is seen as a process of mapping, produced by the mutual orientations of individuals, of feelings and actions into an ensemble. These coalesce through a system of bargaining, exchange, cooperation, compliance; social psychology and sociology have described these under various headings, among them game theory, self-presentation, consensus theory, and communication theory.

Buckley adds that no matter which metaphor is used, all deal with "systems of interlinked components in ongoing developmental process" that produce these emergent phenomena.

ROLES AND INSTITUTIONS

Systems theory must consider large-scale institutions. Buckley finds sociological theory unsatisfactory with respect to this problem. The leap from "micro-processes" to large-scale institutions is not successfully made. Discussions usually are based on the assumption that institutions maintain themselves through "normative consensus, legitimate authority, common values, internalization of roles via socialization, and the like." [17] (Here Buckley seems to mean Parsons and Parsonians.) But such terminology is static and loses sight of the dynamics of interpersonal interaction. Buckley points to recent tendencies (in the work of Blau and Homans) that react against this orientation. Two views must be questioned: (1) that "institutional structure is "only" personal association writ large; (2) that institutional structure "molds" the situation of action and personality to the extent that only a minimal residue of choice and decision making remains on the personal level.

In Buckley's view each level interacts and molds the other. He devotes considerable space to a survey of contemporary theorizing to show the emerging systems orientation. On the "macrosocial" level, the system theory model holds as much as on the "micro" level. That is, we see a "system of interacting components," engaged in transactions with both internal and external environments, such that knowledge of the external environment is in some way built in to the system. Buckley draws on the cybernetic model here: The system develops selective mechanisms by which it copes with the variety of the environment.[18]

Thus the "sociocultural system is to be seen as such a complex, adaptive organization of components. . . . It maps the variety of its external environment through science, technology, magic, religion, and its internal milieu through common understanding, symbols, norms, and values."

Sociological terms such as *esteem, prestige, authority, power, leadership,* and the like refer to the mechanisms of social selection that underly group decision making—that is, "the selection of communication content and interaction networks, of the rule-making apparatus, of ecological settings and physical layouts . . . all of which work to channel actions, attitudes, collective behavior, decisions.

Out of all this new social forms and structures are generated, sometimes slowly, sometimes abruptly, sometimes with greater, sometimes with less, conscious purpose. The building up of such institutions is a selection process, leading to some institutions that are legitimate and others that are not. But the final concrete form of such processes is not "determined" by the social system, rather by the ongoing "interactions and accommodations" of the components as they come into contact; they are created by particular selective and perpetuating mechanisms.

Structure is never self-maintaining; a perpetual expenditure of energy is required to maintain its "steady state;" discrepancies and pressures lead to continual remapping and reorganization.

INSTITUTIONALIZATION

The next task, for Buckley, is that of conceptualizing and systematizing the "more detailed mechanisms and processes underlying the social order." Recent contributions to sociological theory are classified by Buckley under three main headings: the structural and categorical approach, the collective behavior approach, and the social-psychological approach. The task for the system theorist is to develop a framework that incorporates all of these, or at least their essential approaches.

Buckley appears to find that whatever is significant in older sociological theory, especially that of Weber, is akin to this approach; Weber especially "flirted with" the underlying psychological and interactional dynamics of institutionalization. Parsons' selective adaptation from Weber was based on the analogy between society and the species, not the organism. All such developments were strivings to account for the emergence of modern economic institutions out of the network of premodern institutions, as in the work of Parsons and Smelser. These were steps in the right direction.

The Collective Behavior Approach to Institutionalization

Many features of modern social institutions are generated, maintained, and rejuvenated by "relatively unstructured collective processes." Thus social unrest leads to collective or mob action, which in turn leads to public discussion, opinion formation, more organized forms of collective decision-making and action, including social movements, political parties, and even revolution. All this in turn culminates in the institutionalization of a new order.

Minimally structured situations often spontaneously generate new norms, values, and symbols; many qualitative changes in economic and political institutions are the cumulative results of largely uncoordinated plans of action developed within organizations.

An institution is an ongoing, circular, systematic process based on feedback or pseudo-feedback processes.

EXCHANGE AND NEGOTIATION MODELS

Exchange theory, bargaining theory, interactionist role theory are all examples, in Buckley's view, of the way in which current theory is moving in the
systems direction. These theories have revived the Hobbesian problem of
order. By adopting an extreme Hobbesian position—seeing man as totally
self-interested, unsocialized, unconstrained by social norms—theorists such as
Coleman show that norms develop as an end-point result of the feedback to
persons and to collective groups, who learn the long-term consequences of
their actions. The Parsonian appeal to norms fails to recognize the large
element of calculation and manipulation in behavior. The problem is to
reconcile the Hobbesian "presocial" level with the facts of interdependence.

An important step is taken in this direction by Homans' theory of institutionalization. Homans sees activities as sustained by either "natural" or by
"contrived" rewards (such as money or social approval). As networks of obligations and ties develop, so does the exchange of rewarding activities.

Institutions, then, which are the rules governing the behavior of many
people, are continuous with elementary social behavior and grow out of it.

Buckley sees Homans' view as more dynamic and descriptively adequate
than are the static normative approaches, for all its defects. Through the
distribution of rewards, institutions may be founded, may maintain themselves in spite of deficiencies, may disintegrate, and may persist in a state of
unresolved malaise and conflict. Whatever its defects, this theory at least
takes into account the dynamic processes at the social-psychological level of
interpersonal transaction.

Blau's views of power and exchange are somewhat akin, expect that Blau
differs from Homans in deriving sociological concepts through an "emergence" point of view: Out of elementary behavioral phenomena emerge intermediate concepts such as status, power, legitimacy, and the rest. Blau also
differs from Homans in limiting his theoretical approach only to rational
actions—that is, actions oriented toward ends that can be achieved only
through interaction with others. He avoids trying to account for all possible
forms of behavior.

SOCIAL-PSYCHOLOGICAL APPROACHES

Social organization is not a determinant of social action but a framework
within which it takes place; the organization and its changes are the product
not of "forces" but of the acting units. Social interaction has a "process"
character, of creating and re-creating meanings and expectations in a succession of situations that are not regularized or standardized. Norms exist, but
are always creatively reaffirmed from day to day.

Role behavior is seen as morphogenic; a formal role is a "skeleton" of rules; as it is played out it may become "built up and consciously validated." The purposes and sentiments of the actors constitute the unifying element in the genesis and maintenance of the role.

Since roles actually leave lattitude in their performance, those filling the roles are left scope for choices and decisions; roles are then internally and externally validated as they are filled by their incumbents, who carry out working compromises between normative frameworks supplied by others, and the requirements of "their own purposes and sentiments."

Thus whether institutional processes are seen as "negotiated order" by some sociologists, or as the resolution of role strain by others, they all exemplify, for Buckley, the recent convergence of views which . . . is seriously challenging the normative-structural point of view.

Thus Buckley places major emphasis on the replacement of older static views of social processes by a stream of newer thought, tending, in all its tributary streams, towards the systems view.

Buckley's concluding argument is in behalf of the systems approach in understanding the theoretical problems associated with social control, legitimacy, and deviance. Conformity and deviance are both "processes" and as such are very much of the same nature. Deviance is a system product emerging out of interpersonal transactions, role strains and their resolutions, bargains, random or trial deviations, as much as is conformity. If certain feedback loops amplify the deviation behaviors, they may be labeled as such and even defined as crime. A self-inforcing process of alienation from the community may occur, along with deviation-reducing pressures.

Social goal-seeking is a concept seen as especially amenable to cybernetic treatment. Governments may set goals, establish administrative machinery for achieving goals, and receive feedback information on goal-achievement or goal-deviation. But in this area sociological theory is as yet undeveloped; here, as well as in the area of taking corrective action, much work has to be done.

POWER, AUTHORITY, LEGITIMACY

After surveying and rejecting the "standard" conceptions of power, authority, and legitimacy to be found in the literature in Weber, McIver, Parsons, Lynd, and others, Buckley offers Peter Blau's recent *Exchange and Power in Social Life*[19] as an illustration of a system concept of power and authority.

A sociocultural system having an optimum degree of stability and flexibility, relatively stable "social-psychological foundations of interpersonal relations and of the cultural meanings that hold group members together, . . . a flexibility of structural relations characterized by the lack of strong barriers to change, along with a certain propensity for reorganizing the current institutional structure should environmental challenges or emerging internal conditions suggest the need" will exhibit the capacity to persist by changing its own structure. At the basis of this process are the systems-cybernetic features that are by now familiar: a source of new "variety" introduced into the system, which refines or revitalizes "the pool of commonly usable information;" maintenance of an "optimum" level of tension in the system, along with basic satisfaction of members' needs; a complete communication network that extends throughout the system making two-way feedback loops possible; a decision-making center that is appropriately sensitive to both internal and external information and is able to "learn" and change its goals and values; and finally, "effective mechanisms for preserving and propagating those meanings, symbol systems, and information sets that have, for the moment, passed the tests of truth, goodness, and beauty; and this newly structured variety becomes the basis of the sociocultural framework within which the next round of adaptive process occurs." [20]

Such a model is complex, but superior to the equilibrium or functionalistic model, though perhaps less comforting; nevertheless, the challenge has been accepted, and Buckley foresees a revision of sociological theory in the systems-cybernetic direction.

Buckley's work was published in 1967; since then sociology appears to have made comparatively little use of the systems approach.[21] Talcott Parsons' reapproachment with systems theory may be an indicator of future changes, but as of this writing systems theory in the social sciences has been more prominent in political science than in sociology proper.

SYSTEMS THEORY IN POLITICAL SCIENCE

Karl Deutsch of Yale University and David Easton of the University of Chicago are the leading representatives of the systems approach in political science. Deutsch's primary missionary work is *The Nerves of Government*, to which he has added a number of articles in scholarly journals.[22] In addition, his *Nationalism and Social Communication* attempts to apply cybernetics and communication theory to the problem of nationalism and national consciousness.

THE NERVES OF GOVERNMENT

The major portion of this work is a statement of systems theoretical concepts, coupled with a declaration of their relevance to political science. In the familiar manner of a programmatic manifesto, Deutsch begins by describing the

history of sociopolitical thought before the emergence of the cybernetic-sys-
tems theory approach, and the outdated models underlying them, among
them the mechanistic and organismic. Adopting the language of methodology, he tells us that models must have the properties of economy, relevance,
the ability to generate predictions, organizing power, and combinatorial richness. After surveying the familiar models of Hobbes, Locke, Montesquieu,
Edmund Burke, and other "classical" figures, he moves to describe those of
Weber, Parsons, and Merton. Most favorably mentioned, however, are still
more recent models, especially von Neumann and Morgenstern's game theory and Thomas C. Schelling's *Strategy of Conflict*. All this is by way of preparation, however, for his advocacy of the "newest" model, cybernetics. The cybernetic concepts presented by Deutsch are by now familiar; what little is
new is the manner in which he adapts them to the vocabularly of political
science. But to do this a number of steps must first be taken. Deutsch, like
Ashby, tries to "cyberneticize" subjective and psychological processes; that is,
the concepts of consciousness and will, memory and recognition, must be
translated into cybernetic terms.

Consciousness may be defined, as a first approximation, . . . as a collection of internal
feedbacks of secondary messages. *Secondary messages* are messages about changes in the
state of parts of the system, that is, about primary messages. *Primary messages* are those
that move through the system in consequence of its interaction with the outside world.
Any secondary message or combination of messages, however, may in turn serve as a
primary message, in that a further secondary message may be attached to any combination of primary messages or to other secondary messages or their combinations, up
to any level of regress.[23]

Similarly, for the will:

Will, . . . may be tentatively defined, in any sufficiently complex net, nervous system,
or social group, as the set of internal labels attached to various stages of certain
channels within the net, which are represented by these labels as relatively unchanging. . . .
In other words, *will* may be called the set of *internally labelled decisions and anticipated
results, proposed by the application of data from the system's past and by the blocking of incompatible
impulses or data from the system's present or future.* Since the net cannot foretell with certainty either the outcome of the subsequent trains of its own internal messages and switching orders, or the outcome of its own efforts to inhibit information incompatible with
the "willed" result, it knows only what it "will do," not what it "shall do." It may
"know its mind," but it cannot know with certainty whether or when it will change
it.[24]

The second step involves attributing "will" and "consciousness" as defined above to "the political system." Societies receive information, "learn," make decisions, adapt to their environments by means of effector units, and evolve.

If we think of an ethnic or cultural community as a network of communication channels, and of a state or a political system as a network of such channels and of chains of command, we can measure the "integration" of individuals in a people by their ability to receive and transmit information on wide ranges of different topics with relatively little delay or loss of relevant detail. . . .
Similarly, we can measure the speed and accuracy with which political information or commands are transmitted, and the extent to which the patterns contained in the command are still recognizable in the patterns of the action that are supposed to form its execution. . . .

If many studies of politics have stressed *power,* or enforcement, it should now be added that information precedes compulsion. It is impossible to enforce any command unless the enforcing agency knows against whom the enforcement is to be directed. . . .[25]

Similarly, information must precede compliance in that one must understand a command in order to obey it. This suggests, for Deutsch, an important area for political research: the levels on which communications operate most effectively for coordinated political action. He illustrates this with reference to military structures: There are too few generals and too many privates for meaningful and direct channels of communication (and action) between them; similarly, there are too many sergeants and lieutenants for them to organize effectively; but colonels appear to be most favorably placed both in numbers and in hierarchical position to engage in political intrigue, and they are in fact prominent in this respect. In any event, regardless of specific illustrations, Deutsch moves explicitly toward personalizing and psychologizing the political system. Deutsch even finds it necessary to find cybernetic terms for ultimate values, among them *autonomy, integrity, meaning,* and *spirit,* apparently because they would have no foundations outside of this vocabulary:

A society or community that is to steer itself must continue to receive a full flow of three kinds of information: first, information about the world outside; second, information about the past, with a wide range of recall and recombination; and third, information about itself and its own parts. Let any one of these three streams be long interrupted, such as by oppression and secrecy, and the society becomes an automaton, a walking corpse. It loses control over its own behavior, not only for some of its parts, but also eventually at its very top.[26]

Deutsch defines the *self* as placed "at the location of the feedback circuits of the relatively highest type. . . ." "Selfhood . . . appears not as a static property but as the functioning of particular sets of channels in a communications system. Self-determination may increase with the increase in the number, effectiveness or organization, and level of type of these sets of channels." Deutsch then defines *integrity* as "the unimpaired functioning of the *facilities that carry the processes of self-determination*. The 'integrity' of any self-steering system can be impaired either by imposing a nonautonomous change on some of its channels—such as by curring a channel or disrupting its connections—or by leaving its channels intact, but forcing such traffic loads on some of them as to disrupt their functioning." Defense of integrity is seen by him as defense of the capacity to learn, which includes "the structure of the personality we have acquired."

Dignity is nondisruptive learning. Integrity is undisrupted or unimpaired inner learning equipment. . . . To ensure continued self-determination, integrity and dignity are not enough. In the language of religion, pride may mean death where a change of heart may mean survival. In less exalted terms, the best way to strengthen a communications system against the impact of large external changes may well consist in enriching its internal structure and its range of possible new configurations.[27]

This switch from the language of theology to that of cybernetics is rather puzzling, because it does not appear to add anything to the problems of the moral theology of integrity, free-will, and self-determination which have been argued for centuries. To describe these problems in terms of feedback circuitry adds nothing to their resolution; furthermore, to discuss the problems of integrity, autonomy, and dignity, without reference to the experience of both outer and inner pressures, to the experience of domination, to the facts that socialization inescapably involves the learning of self-limiting and often self-damaging forms of thought and behavior, as well as the learning of their opposites, is simply to develop imagery without discussing meanings. Aside from a certain unctuous moral tone, nothing else seems to be added.

Deutsch now feels able to define mind and creativity. *Mind* is provisionally defined:

. . . any self-sustaining physical process that includes the nine operations of selecting, abstracting, communicating, storing, subdividing, recalling, recombining, critically recognizing, and reapplying items of information. . . . Such a physical process may include the production of novelty and of initiative. If information from events in the outside world is abstracted and stored analytically, . . . then these may be separately recalled and recombined into new patterns that did not exist in the outside world. . . . To complete the production of *novelty*, this new combination of old elements must itself be matched or abstracted in the mind: a new image or symbol must be stored, *pertaining to the new pattern as a whole, regardless of its earlier combinatorial origin*.[28]

This is creativity.

And at last we are told what *spirit* is.

Spirit denotes second-order value. It is the set of preferences about sets of preferences. A man, a people, or an epoch are among other things also systems; the *spirit* of a man, or a people, or an epoch is the configuration to which their value systems are patterned or operated. Spirit is related to values as strategy to tactics or as policy to operations. A change in "spirit" means, therefore, a strategic change in the patterns of behavior. And, under suitable conditions, such a change can be communicated.[29]

Having "created" and personalized the "system," Deutsch moves to the positing of this system in the political sphere. Political systems too have spirit and will; they struggle to survive with greater or lesser probability of success; they may be *self-destroying* (apt to break down even under favorable conditions), *nonviable* (unlikely to survive under most environmental conditions, though not self-destroying), viable (likely to survive over a limited range of environmental conditions), *self-developing or self-enhancing,* "which are able to increase their probability of survival and their ranges of possible action over an increasing variety of environments." Social organizations, though not biological organisms, are like them in exhibiting features of "health" and "evolutionary progress." The politics of power and growth can then be discussed in terms of manpower, economic growth, operational reserves to meet new stresses or challenges, and self-determination—that is, "the growth of resources and functions that bear on social cohesion," internal communications, and in the growth of the sterling performance of the system. Implied in this is the ability to develop "strategic simplifications," without which the increasing number and complexity of messages would become unmanageable. Such strategic simplifications are exemplified in the invention of the written alphabet and finally of printing, the replacement of telegraph wires by radio messages, of trolley tracks by rubber tires, as well as the advance from the elaborate models of "Ptolemy and Copernicus to the simpler and more general formulations of Newton and Einstein." The viability of a system involves its ability to arrive at goals once chosen and to change goals when appropriate. To be sure, a human organization is not an anthill, and true growth consists in an interplay between the growth of the organization and the growth of individuals and the subgroups within it. "In this sense, the growth of human organizations is always the growth of several levels of autonomous systems, and the autonomous growth and enhanced self-determination of individuals is one of its touchstones."[30]   The task of politics is accelerating needed innovations, though Deutsch is not specific about which innovations are needed, and when. Politics can be used to rigidify social systems and to accelerate change. Western politics have developed three techniques for bringing about change: majority rule, the protection of minorities, and the institutionalization of dissent. These techniques have made possible a wide range of resources for "rapid social learning and innovation," through which "mankind in its various subdivisions, still organized in states, can adapt more quickly to the dangerous but hopeful tasks of growing up." Deutsch appends an impressively cybernetic diagram as foreign policy model, consisting of squares and triangles connected by a network of arrows. Thus arrows representing stimuli enter boxes labeled Foreign Input (receptors), Domestic Input (receptors), which are connected to a Screen of Selective Attention, which in turn is connected to a number of other boxes and triangles, among them: Selective Recall, Selective Memory, Current Memory, Deeply Stored Memory, Screen of Acceptable Recalls, Tentative Decisions, Screen of Repression from Consciousness, Confrontation and Simultaneous Inspection of Abridged Secondary Symbols (consciousness), Screen of Acceptable and Feasible Policies, Final Decisions, and, outside the maze, Foreign Output (effectors) and Domestic Output (effectors). The diagram is accompanied by a list of statements explaining the components of the diagram—for example, "F_1 Feedback Information about the results of foreign policy actions, C_1 Combined selected data and acceptable memories, moving toward final decision (e.g., 'action papers')."

The criticisms of the approach are obvious and have been made before, especially with reference to cybernetic theory itself: that it represents a mechanistic "engineering" approach to politics,[31] that it is based on specious analogies, that it is excessively formal or rationalistic. To these we would add the criticism made earlier only by implication: that Deutsch's work, like the major part of systems theory, bears little or no empirical reference whatever, solves no theoretical or practical research problems, and appears to be little more than an exercise in vocabulary. Still further, it must be said that the "political system" referred to throughout is so abstract and vague that no one can recognize any specific political reality in it—American, European, or whatever.[32]

DAVID EASTON

The programmatic statements of David Easton are in close agreement with those of Karl Deutsch, whose work Easton regards as "subtle and penetrating."[33] In fact, they add little to what has been said. Nevertheless, the evolution of Easton's ideas did produce some minor academic developments that are not fully identical with systems theory in its mature form; these developments include an extended flirtation with the movement known as behavioral science. For a time, Easton appeared to be moving in this direction, but later, without explicitly disavowing the behavioral science orientation, he adopted his own synthesis of operations research, cybernetics, and general systems theory.

Easton's work is described by him as part of a long series, guided by a continuity of purpose. His thought is embodied in three books spanning a 12- to 15-year period: *The Political System* (first edition, 1953); *A Framework for Political Analysis* (1965) continues the program; and *A Systems Analysis of Political Life* (also 1965) completes his program, at least so far.[34]

Easton originally described his first work *The Political System* as a document of "behavioralism," and in fact a document that helped establish the "behavioral revolution" in political science: "Although those of us who have struggled to consummate the behavioral revolution have hailed it as perhaps the most significant transformation in the whole history of political science, it did not herald the millenium. It has now become evident that behavioralism has been but one stage, although a profound and vital one, in the continuing evolution of our discipline."[35]

The work follows what by now appears to be an obligatory format for all programmatic documents: The [then] current mood of the world is described as one of malaise; the appearance of atomic weapons has led to a distrust of science, even of political science; the condition of American political science is examined and found wanting, and exhorted to rise to meet new challenges. The concepts and methods of the past are not adequate to the needs of the present. Interwoven with this argument is a survey of political concepts since Plato and Aristotle, through Macchiavelli and Hobbes, to such current or nearly current figures as Merriam, Wirth, Gosnell, and Lasswell. Standard concepts such as the state, power, elites, and the like are no longer sufficient. New ones must be added. Easton appears to propose a new concept as central to political science: "the authoritative allocation of values for a society."[36] Such a concept enables us to understand three central concepts: policy, authority, and society. By policy is meant the decision process by which "certain things are denied to some people and made accessible to others"—a web of decisions and actions that allocates values. The political scientist must study the entire process of policymaking, both formal and informal. But this leads beyond the narrowly conceived political structure to a broader concern with how an entire society allocates values, as well as to a concern with the problem of the nature of authority. Thus political science research must concern itself with the study of policy activities.

Once the major concepts are delineated, we must next consider what classes of data should be incorporated into the theory. An important step must be taken in freeing political science from overemphasis on juristic and purely legal matters, in favor of examining more closely how people in political situations actually behave and the total structure of the group situations within which they act. Situational data are the basic materials of political science; there are three kinds: the physical environment, the nonhuman organic environment, and the social environment (at which point Easton veers close to a systems approach; the emphasis is on the importance of the physical environment, ecology, and social environment). Behavioral research will seek to encompass the methods and concepts of all social sciences, among them sociology and psychology, by studying "situational behavior." With the establishment of this general framework, political science will be well-positioned for making useful contributions to the understanding of political life. But theoretical research must be placed on clear moral foundations; values may underlie research but should not influence it, and the same is true of theory. The only way for the researcher/theorist to control this influence is by the development of self-critical clarity regarding the moral roots of his work.

Next, Easton notes that political theory is in a general state of decline, being sunk in one or another form of historicism; Easton, in a step that bring him close to his later systems thinking, proposes the concept of equilibrium analysis as fruitful for the discipline, fruitful even in its inadequacies: "The idea of a general equilibrium implicit in so much empirical work in political science, . . . can help to perpetuate the notion that political activity is part of an empirical system and of a process of change through time. These are insights which future attempts at theory construction can scarcely neglect." The concept of equilibrium is implicit in the works of a number of political scientists who might be surprised to know it was there; this lack of awareness is due to the traditional emphasis in political theory to the history of moral ideas and a corresponding neglect of either empirical description or general theory. The development of a systematic theory remains the unfinished task of political science.

The Political System, then, is more of a call for a systems theory than it is systems theory proper. Only in the two later works and in his additions to *The Political System* for its second edition does Easton explicitly embrace modern systems theory.[37]

Easton's second work *A Framework for Political Analysis* explicitly embraces systems analysis with certain qualifications. Although his notion of systems thinking is related to the systems concept in sociology and economics, he emphasizes his relationship to the "communications sciences," which have gone far beyond the older systems approaches.[38]

He regards this approach as a further step in the development of "behavioral research," one that gives it more powerful conceptual tools, and which will hopefully accomplish the integration of the social sciences.

The conceptual framework offered by Easton is extremely close to systems thinking in the cybernetic-communications sense, to operations research, and to von Bertalanffy's general system theory; it can be described as a blend of the three. Political life is seen as a *system:* Systems exchange influences with *environments;* systems exhibit *responses* to internal and environmental stresses; they are able to do so due to the presence of *feedback.*

In general, systems analysis, as I shall conceive it, takes its departure from the notion of political life as a boundary-maintaining set of interactions imbedded in and surrounded by other social systems to the influence of which it is constantly exposed. As such, it is helpful to interpret political phenomena as constituting an open system, one that must cope with the problems generated by its exposure to influences from these environmental systems. If a system of this kind is to persist through time, it must obtain adequate feedback about its past performances, and it must take measures that regulate its future behavior.[39]

But systems need not be regarded only or exclusively as "natural" systems; artificially constructed (constructive) systems may prove equally or even more fruitful. They need not be reified.

But not all systems active in society are political systems. How are these distinguished? The political system is "the most inclusive system of behavior in a society for the authoritative allocation of values." To distinguish political roles from all other social roles, *person* refers to an individual in his totality, *member* to an individual acting as part of the political system. Political systems have boundaries, but these are more difficult to conceive than are the boundaries of biological entities; they are not to be confused with geographical boundaries or borders, though at times such borders may coincide with system boundaries; true system boundaries are often a matter for empirical determination. To some extent this depends upon the degree to which political roles are differentiated from other social roles in any given society. Easton too resorts to diagrams, with one box labeled The Political System, connected by arrows to other boxes labeled The Intrasocietal Environment, The Extrasocietal Environment, which are in turn connected to The Total Environment, which comprises many systems, among them ecological, personality, biological, social, and other systems.

Systems thinking with respect to the persistence of systems is adopted bodily by Easton; political systems are subject to stress, both internal and external; systems may disappear or change almost out of recognition; other systems "persist," primarily by means of change; the systems analysis theory of persistence is not to be confused with theories of self-maintenance or theories of equilibrium. Here Easton relates himself to cybernetics, borrowing from Ashby the concept of essential variables that may or may not be pushed beyond their "normal" range by stress. If the variables do move beyond certain critical limits, the system may change, collapse, or disappear. The essential variables in a political system are identical with "the capacity of the

system to allocate values for the society and to assure their acceptance."
Exactly what the acceptable range of an essential variable is may not always
be easy to determine, but it is in principle determinable. It suffices to keep in
mind that essential variables have critical points though these are unknown.

An important related point is to determine the particular kinds of regulative responses "that characterize all systems." This is reserved for a succeed-
ing volume.

Easton is vulnerable to criticism, as is all systems theory, in his argument
that systems have essential variables that must be kept within acceptable
limits; the argument, if not tautological or question-begging, appears to have
no empirical referents, and cannot be falsified. Furthermore, the responses he
describes appear to be drawn from political history rather than from the
substance of the system theory. The theory appears to follow events, rather
than to order or illuminate them.

Similar objections apply to his view that where one system may disappear
through failure to respond appropriately to stress, another may cope by drawing upon its reserve of past experience. He suggests further that systems approaches need categories of analysis that will make it possible for us to understand the modes of response that are available to political systems. No doubt,
but they are not yet forthcoming.

For the rest, Easton adopts the vocabulary of systems theory, including the
notions of feedback, input overload, outputs, exchanges, and the like, occa-
sionally illustrating these concepts in political terms. He concludes as though
systems terminology offers the founding charter for behavioral research.

Behavioral research represents for the first time a commitment to the broad and
essential requirements of scientific knowledge: the search for criteria that within the
scientific framework will permit the investigator to test for the relevance of empirical
data and at the same time will offer some hope of providing a richer understanding of
the phenomena central to his interest. This is uniquely the task of theory and it is to
its construction that modern behavioral science has slowly been guiding our footsteps,
so slowly indeed that only in the last decade has it become even faintly visible.[40]

He adds that the computerization of research opens up magnificent new
vistas of inquiry. At last political science can become rigorous.

Easton remains on the level of metatheory in *A Systems Analysis of Political
Life*. Again, the vocabulary of systems theory is developed and extended in
greater detail, but the work remains an exercise in vocabulary. No real-life
political system available for analysis on earth is examined, though much
political material is cited here and there by way of illustration. Political
systems are described as linked to other systems—social, economic, ecological,
and the like. There are linkage variables between systems; there are inputs

and outputs; there are feedback loops; decisions are diagrammed; but these are all prolegomena; here a "structure for the analysis of political systems" will be constructed.[41] But this is prolegomenon to prolegomenon, because the third volume consists entirely of more conceptualizing within the framework of a systems terminology. Within his own framework, Easton fails to maintain even minimal discipline of thought, at times defining "system" as any set of arbitrarily chosen variables (provided they are "interesting"),[42] at other times reifying the system as a naturally existent reality; systems then persist, succeed, fail, take appropriate or inappropriate steps, develop responses, and the like. What is developed is not a theory in any sense of the term, but a vast image or metaphor in which the already existent picture of political life is converted into systems terminology, without adding anything new by way of terminology, concepts, insights, or findings. Social "wants" are converted to "demands"; demands are conveyed to the political system by "channels of information flow"; adaptive responses of one or another kind are developed by the system, all of which "arrives" at such new and original concepts as legitimacy, pressure groups, elite ties, charisma, and ideology. Systems vocabulary even discovers for us that ideologies have two aspects—expressive, and instrumental[43]—that regime responses to stresses may be adequate or inadequate, and may increase or decrease social cohesion.

Thus the vocabulary of systems analysis in political science, as developed by Easton and Deutsch, appears as a vast and elaborate detour by means of which we arrive at precisely the concepts that are the heritage of political science and sociology bequeathed by such earlier theorists as Parsons, Weber, Mosca, Michels, and others. If the conceptual apparatus borrowed from cybernetics is subtracted from this material, virtually nothing remains but a thin stucco, composed for the most part of surveys of earlier political theory. The primary failure of this school of theorizing, besides its substantive sterility, is its failure to deliver according to its own claims. Truly fruitful scientific theory—at least on the scientistic model proclaimed by the systems thinkers—consists of the manipulation of symbolic or conceptual systems to achieve results that do not always accord with common sense, but that gradually filter down and become common sense. What was once abstract and esoteric becomes something "obvious" to all; the roundness of the earth is now something that is naively perceived, whereas once it was an abstract theory that contradicted naive realism. But systems theorists such as Deutsch and Easton make no true attempt at such a use of systems concepts; at all times, they appear to keep one corner of their eyes on already existent concepts to arrive carefully at images that coincide with such concepts, dressed up as they are in new terminology. But this new terminology is their only contribution.[44]

NOTES

1. Systems theory in anthropology is not discussed in this chapter. The primary reason is that systems theory in the modern cybernetics sense has had relatively little impact on anthropology, although certain closely related streams of thought—among them organicism and evolutionism—have. A thorough discussion of what might be termed "systems-like thinking" in anthropology would require a book-length study and so is beyond the scope of the present work. Nevertheless a few observations can be made here.

   Perhaps the full adoption of cybernetic systems theory in anthropology has been delayed by the fact that anthropologists have already developed a number of abstract formalisms of their own. One of these, for example, is the elaboration by computer of kinship structures. Another is the long and elaborate debate between the evolutionists and the diffusionists, a topic beyond the scope of this study.

   Radcliffe-Brown is the anthropologist whose works come closest to modern systems theory in the sense of General System Theory, but even his main testament, *A Natural Science of Society*, predates the movement of von Bertalanffy and the cyberneticists. Nevertheless, the affinities are very great. Gregory Bateson's *Steps to an Ecology of Mind* appears to be the only major attempt by an anthropologist to incorporate information theory and cybernetics into anthropology, but this work is of too recent vintage (1972, though some sections were written and published much earlier) to have established a major movement of thought in the field.

   An adequate discussion of these issues would extend the present work unduly. The following sources provide some discussion of them, and in general of the controversies surrounding the weaknesses of organic and evolutionary ways of thinking about social and historical matters: Kenneth Bock, *The Acceptance of Histories* (Berkeley: University of California Press, 1956); Robert Nisbet, *Social Change and History* (New York: Oxford University Press, 1968); Margaret T. Hodgen: *Early Anthropology in the 16th and 17th Century* (Philadelphia: University of Pennsylvania Press, 1964; reprint, 1971).

   One of the major weaknesses of all systems theorists has been their uncritical acceptance and transfer to social matters of an organic and evolutionistic manner of thinking. The concept of evolution, both in its early Darwinian sense and in the later revisions of this by contemporary biologists, is fraught with problems, which I consider fatal to an evolutionary point of view. The systems theorists, in striving to be "scientific" and "up to date," have incorporated perhaps the most archaic and least tenable feature of nineteenth-century thought. Kenneth Bock's article, "Evolution, Function and Change," *American Sociological Review,* (Spring 1963) Vol. V, pp. 229–237, provides a critique of such thinking in Parson's work, but a much broader critique of evolutionary thinking is possible. For one such, which, however, does not touch upon the social sciences, but whose critique is wholly applicable to the systems theorists, see the important work of Norman Macbeth, *Darwin Retried* (New York: Delta Books, 1971).

2. Talcott Parsons, *Societies—Evolutionary and Comparative Perspectives* (Englewood Cliffs, N.J.: Prentice-Hall, 1966), p. 1. Parsons indicates (p. 6) that it is really only the first part of a two-part work.

3. *Ibid.,*

4. *Ibid.,* pp. 8–10. Parsons' footnote (p. 9) is also relevant in this respect: "The theory of cybernetics was first developed by Norbert Wiener in *Cybernetics* . . . and was applied to social problems in his *The Human Use of Human Beings* . . . A good introductory statement for the social scientists will be found in Karl Deutsch, *The Nerves of Government.* . . ."

5. Talcott Parsons, *The System of Modern Societies* (Englewood Cliffs, N.J.; Prentice-Hall, 1970), p. 2. Parsons indicates that these two books were originally intended as one but were divided because of space limitations.

6. *Ibid.*, pp. 5–7. Parsons makes no mention of Ludwig von Bertalanffy in connection with "open systems that interchange inputs and outputs with their environments."

7. For the rest, we assume that Parsons' thought is well enough known both from primary and secondary sources to need no further exposition here. It might be noted, however, that Parsons' organic evolutionism has had no predictive value whenever it has been applied empirically, a failure reminiscent of the failure of systems theory to predict urban developments. Thus Parsons' predictions with respect to race relations do not appear to have been borne out. In this connection, see Parsons' essay, "Full Citizenship for the Negro?" in Talcott Parsons and Kenneth Clark, *The Negro American* (Boston: Houghton Mifflin, 1968) and the critique of it in Stanford M. Lyman, *The Black American in Sociological Thought: a Failure of Perspective* (New York: Capricorn, 1972), Ch. 6.

8. Walter Buckley: *Sociology and Modern Systems Theory* (Englewood Cliffs, N.J.: Prentice-Hall, 1967). Walter Buckley, Ed., *Modern Systems Research for the Behavioral Scientist—a Source-book* (Chicago: Aldine, 1968).

9. *Ibid.*, p. 9.

10. *Ibid.*, p. 14.

11. *Ibid.*, p. 14.

12. Parsons, *Societies*, p. 21.

13. Buckley, *Sociology and Modern Systems Theory*, p. 18.

14. Buckley, *Sociology and Modern Systems Theory*, p. 18, p. 36. The discussion above summarizes Buckley's own, pp. 23 ff.

15. *Ibid.*, p. 39.

16. *Ibid.*, Ch. 4, especially pp. 94ff.

17. *Ibid.*, p. 127.

18. *Ibid.*, p. 128.

19. Peter Blau, *Exchange and Power in Social Life* (New York: Wiley, 1964).

20. Buckley, *op. cit.*, Ch. 6, passim; see especially pp. 206–207.

21. Two exceptions: Joseph H. Monane, *A Sociology of Human Systems* (New York: Appleton-Century-Crofts, 1967) and Orrin E. Klapp, *Models of Social Order* (Palo Alto, Calif.: National Press, 1970) are sociology textbooks oriented toward the systems approach, but they are primers. No substantive work appears to have been done at the "frontiers" of sociological theory.

22. Karl W. Deutsch: *The Nerves of Government—Models of Political Communication and Control* (with a new introduction) (New York: Free Press, 1960). Originally published 1963; *Nationalism and Social Communication* (Cambridge, Mass.: M.I.T. Press, 1953).

23. Deutsch, *Nerves of Government*, p. 98.

24. *Ibid.*, p. 105.

25. *Ibid.*, pp. 150–151.

26. *Ibid.*, p. 129.

27. *Ibid.*, pp. 131–133.

28. *Ibid.*, p. 133. Author's emphasis throughout.

29. *Ibid.*, p. 140.

30. *Ibid.*, p. 253.

31. For a more sympathetic description of Deutsch's work, written by a systems theorist, see, Oran R. Young, *Systems of Political Science*, Ch. 4, "Approaches from Communications Theory and Cybernetics" (Englewood Cliffs, N.J.: Prentice-Hall, 1968). Young takes notice of these criticisms (pp. 59ff), but appears to give them little weight. See also W. J. M. MacKenzie, *Politics and Social Science* (Baltimore, Md.: Penguin, 1967).

32. *The Nerves of Government* appears to represent Deutsch's attempt to create a definitive theoretical statement of cybernetic-system theory for political science. The earlier development of his

ideas may be traced in a number of articles indicated below, these add nothing new to his argument, or indeed to the formulations of von Bertalanffy and the cyberneticians. His earlier work includes the following: "Communication Theory and Social Science," *American Journal of OrthoPsychiatry,* Vol. 22 pp. 469–483; "On Communication Models in the Social Sciences," *Public Opinion Quarterly,* Vol. 16 pp. 356–380. (1952); "Mechanism, Organism and Society," *Philosophy of Science,* Vol. 18, No. 3 (July 1951), pp. 230–252; "Mechanism, Teleology, and Mind." *Philosophy and Phemonenological Research,* Vol. 12, No. 2 (December 1951), pp. 185–222.

These articles are noteworthy primarily for the extent to which they repeat the same ideas; in essence, the same article was written four or five times.

33. David Easton, *A Framework for Political Analysis* Englewood Cliffs, N.J.: Prentice-Hall, 1965), p. 19n.

34. David Easton, *The Political System—an Inquiry into the State of Political Science,* 2nd ed. (New York: Knopf, 1971; first edition 1953); *A Framework for Political Analysis* (Englewood Cliffs, N.J.: Prentice-Hall, 1965); *A Systems Analysis of Political Life* (New York: Wiley, 1965).

35. "Preface to Second Edition," *The Political System,* p. xiii.

36. *Ibid.,* pp. 129ff.

37. Easton dates his first approach to systems theory proper as occurring in 1955 when he wrote his "An Approach to the Analysis of Political Systems," *World Politics* (1957), pp. 383–400. Nevertheless, his elaborations of the approach do not appear under the middle 1960s.

38. Easton, *Framework,* p. xi.

39. *Ibid.,* pp. 24–25.

40. *Ibid.,* p. 133.

41. Easton, *Systems Analysis,* p. 33.

42. *Ibid.,* p. 21.

43. *Ibid.,* pp. 296ff.

44. Easton's flirtation with "behavioral science" indicates the existence of yet another "movement" within the social sciences. In *A Framework for Political Analysis* (pp. 4ff), he appears to regard this as a new development in social science. The movement can be dismissed as bureaucratic positivism, abstract, formalist, quantitative in character. It seeks to transcend the divisions between psychology, sociology, economics, political science, and to encompass them all. Easton appears to have distanced himself from this movement, which in some respects established its charter prematurely, without relating itself to systems theory. The movement has now attempted to remedy this oversight, and has produced cybernetic documents of its own. Amitai Etzioni's *The Active Society* (New York: Free Press, 1968) and Warren Breed's primer, *The Self-Guiding Society Based on the Active Society by Amitai Etzioni* (New York: Free Press, 1971) are representative documents. These works add nothing to what has been said above, except to note that in them the self-serving character of the theories becomes even more obvious. ". . . society can be self-guiding as its elites probe the creation and exploitation of new options," Breed tells us (p. 19). Here, society seen as a system manipulated by the systems man from a point exterior to and superior to that system, becomes an explicit ideology. If one may speak of "vulgar Marxism" as distinguished from what Marx actually thought and wrote, so one could here speak of "vulgar scientism," if systems theory itself were not already a caricature.

Systems Theory as Ideology

The Migration of
Systems Theory

From its origins, systems thinking has spread to a number of fields and has even begun to assume the character of a popular fad. On the professional level systems advocates have appeared in fields such as psychiatry and psychoanalysis, social work, organizational theory, and urban planning; on the level of popular culture the "ecologists," population planners, and the opponents of economic growth have all begun to speak of the world system in cybernetic terms.

But as systems theory "migrates" into these fields, it acquires new meanings and to some extent loses old ones, while retaining the besetting vices of system theorists. The term *"system,"* as it moves into new territories, becomes ever more loosely defined and impressionistic in usage, sometimes being used in von Bertalanffy"s sense of an open system, sometimes in the cybernetic-communications theory sense, sometimes (e.g., in operations research) to mean simulation models, and often in a blend of all these.

The description that follows must, therefore, lose some precision, as these meanings become confounded in the stream of popular ideology. But the "besetting vices" referred to above remain unchanged—a weakness for programmatic statements coupled with a scarcity of concrete results, a fondness for abstract schematic formulae and diagrams having little practical reference; a fundamental begging of questions that takes the form of an unstated and presumably invisible shift from concrete world "systems" in their fullness and complexity to closed formal models based on convenient "simplifying assumptions," a shift we are not expected to notice; and finally, the absence of concrete work done beyond the refinement of the system itself.

PSYCHOLOGY, PSYCHIATRY, PSYCHOANALYSIS

Some of the first formulations of the systems approach did in fact occur within the field of psychology. Wolfgang Koehler and Andras Angyal, both philosophical psychologists, were important early influences upon the movement. Koehler, in his *The Place of Values in the World of Fact*,[1] offered a psychological formulation of open and closed personality systems, in which he argued against mechanical equilibrium, basing his reasoning in part upon Cannon's work. Koehler argued against "the machine principle" in psychology in a manner closely related to von Bertalanffy's classical presentation. Similar in character is Angyal's *Foundations for a Science of Personality*.[2] Like Koehler, Angyal worked within the tradition of the gestalt psychologists; like him, he argued for "a broad theoretical frame of reference for the integration of the manifold aspects of human nature and behavior into a unified picture of man"; his approach was to be "holistic" (adapting the term from J. C. Smuts' *Holism and Evolution*). This was to be the organizing principle of psychology; the holistic approach to personality would develop more "specific and positive formulations"; would develop a "definite method of synthesis" according to definite rules; further, it would not borrow concepts from physiology or psychology, but would develop its own concepts.[3]

For Angyal, psychiatry has developed no theory of personality; only a holistic approach promises to achieve one. Thus organism and environment are not to be seen as separate structures, but as a totality in interaction; the dynamics of the biosphere are the source of tensions, demands, and values. But individuals must be linked to others, and to cultures, in ways that extent the limits of personality beyond the biological organism. At all levels—biological, psychological, and social-psychological—system-formation is at work (*system* here means "holistic organization").[4] "In aggregates it is significant that the parts are added; in a system it is significant that the parts are arranged ."[5] Thus the main problem for a science of personality is that of discovering how integrated personalities are formed and how this integration is disturbed or malformed. This problem, of the integration of personality, is probably the most important and the most difficult chapter in the study of personality. Pressure, lack of coherence, mutual invastion (system interference), are among the factors that disturb or hinder personality integration. The entire course of life is to be seen as a gestalt, as a process of gestalt-formation throughout the various stages of life, which themselves are organizing gestalts.

Koehler and Angyal's work is generally regarded by systems thinkers as "pioneer" work, the work of "forerunners." Their work was closely related to the biologism of Uexkull, von Bertalanffy, and Sommerhoff, among others. In view of their relatively early appearance and their relation to von Bertalanffy, it cannot be said that their work represents a "migration" into psychology of systems thinking; rather, they contributed to the foundations of systems thinking in its biologistic form. They may be regarded as being among the founders of the movement, to which psychology contributed from the beginning. But the systems movement within psychology was reinforced by later

waves of systems thinking that were emerging from cybernetics and information theory.[6]

Gordon Allport's 1960 article "The Open System in Personality Theory"[7]
is a landmark in systems theory for psychology. Allport bases his framework
directly on von Bertalanffy and hopes to show that it can incorporate most of
the results of the (then) contemporary schools of psychology. His four criteria
for the definition of open systems are, he suggests, sufficient to encompass
psychological theory. The first criterion, material and energy exchange, encompasses stimulus-response theory; the second, steady-state maintenance,
encompasses all that personality theory has said about equilibrium—tension-
reduction, ego defense, and the like; the third criterion, is the tendency of
systems (in this instance human personality) "to go beyond steady states and
to strive for an enhancement and elaboration of internal order even at the
cost of considerable disequilibrium." Allport cites such psychologists as Mc-
Dougall's "proactive sentiment of self-regard," Goldstein's conception of self-
actualization, Maslow on human growth, Jung on individuation, Erikson's
search for identity, among others. The fourth criterion refers to the capacity
of a system within an environment to form a higher-order system of systems
and environment and to exhibit different characteristics in different environments. The psychological dimension of this is the desire to escape the conception of personality as something "integumented, as residing within the skin."
Lewin's field theory, the theories of Gardner Murphy and of social psychology all see individual behavior as fitting into and as evoked by a variety of
systems. Thus this criterion is transactional: The relations between units,
rather than the individual units, form the system.

The general tenor of Allport's article is reproach to psychologists for an
overly partisan and fragmented point of view; the systems approach provides
the unifying framework.

Psychiatry and Psychoanalysis

Systems theorists have emerged recently within the psychiatric and psychoanalytic professions, as well as within the family therapy movement, itself an
outgrowth of psychoanalysis; and missionary books and articles have appeared here too. Emanuel Peterfreund's *Information, Systems, and Psychoanalysis*[8]
brings the systems message to the psychoanalytic world.

Beginning with a critique of current psychoanalytic theory from the viewpoint of the philosophy of science, Peterfreund suggests that the scientific
status of this theory is dubious. Psychoanalytic theory is centered purely on
"psychological ground," with little or no relation to neurophysiology, biology,
embryology, "and all of evolutionary time. . . . Mind is not viewed as a
manifestation of biological activity; it is viewed as being in control of biologi-

cal activity."[9] Just as the ancient astronomers saw the universe as centered upon the earth, so psychoanalysis "is still attempting to force the world of biology, physiology, and evolutionary time into a world at the center of which is the mind of man. This has been the pitfall of a theory which remains on psychological ground." Psychoanalysis, in remaining in an "anachronistic vitalistic position," has become remote from twentieth-century science. Peterfreund finds psychoanalytic theory especially deficient in basic concepts such as psychic energy and the concept of the ego. Psychoanalysis conceives of psychic energy in terms reminiscent of the vitalistic fluids of early physical scientists; energy "flows," can be "transformed," has directional properties, can be attached to objects, and the like; all of this conceptualization is, for Peterfreund, out of date and hopelessly inadequate both to the problems of psychoanalytic theory and when compared to the modern physicist's concept of energy. "Psychic energy is essentially a vitalistic concept." In addition, he finds the psychoanalytic concept of the ego to be unsatisfactory in that it retains the mind-body dualism. The psychoanalytic concept of psychic energy appears to be the bridging point at which analytic theory switches from mind to body and back again. For Peterfreund, the psychoanalytic concept of the ego is tautological, pseudo-explanatory. The concept can be eliminated from all descriptions of the biological human organism, with no loss of accuracy. The general thesis is that analytic theory is a jumble of unclarified, anthropomorphic concepts. "For example, . . . psychoanalytic theory cannot explain how an ego can gain control of inborn biological apparatuses and functions when no relationship has ever been established between psychic energy and physiological energy."[10]

Psychoanalysis, despite its valid empirical clinical work, is confused on the theoretical level; mind is conceptualized according to three languages: that of physiology and biology, the "language of the psychic apparatus," and "the language of persons." "The concepts of energy, force, structure, function, channels, and so on, are used, but they do not have the meanings generally accepted in the scientific world. The concepts of energy (in its general scientific sense), matter and information, and the theory of evolution have no real place in current psychoanalytic theory." In addition, the mind is viewed anthropomorphically; the ego is meaningless abstraction, and the basic model of analytic theory is "hydrodynamic." Consequently, its scientific standing is quite limited.

The deficiencies in Peterfreund's critique, as so far presented, are obvious. First, with respect to his objection to a theory that attempts to remain on "psychological" ground, the answer is simply that even an argument refuting what he calls anthropomorphic approaches to the mind proceeds by mental operations and would succeed, if it could, by affecting minds; even the determination to eliminate mind would be a mental operation. Second, with respect to his denial of the ego and of ego functions ("mind is not viewed as a manifestation of biological activity; it is viewed as being in control of biological activity"), one wonders what he would make of the phenomenon known as biofeedback. (Individuals can, by pure "thought-operations," influence biological reactions.) The problem is not so simple, and an adequate language for the clarification of this problem does not yet exist.

Peterfreund maintains, and cites Freud himself in support, that psychoanalysis is a science in its primitive early phases, whose theoretical framework and fundamental concepts are incomplete, unclear, and even unsatisfactory. Psychoanalysis, he says, is valuable as an empirical or clinical discipline, but much more questionable as "high-level theory." Unless psychoanalysis accepts the view of modern twentieth-century science, which wisely rejects the concept of the ego and of psychic energy, while accepting the concept of unconscious processes, it has little future.

At this point Peterfreund introduces his model, designed to bring psychoanalytic theory into the twentieth century. It is information systems theory, relying heavily on the logic of computer programming, with some admixture of cybernetic theory in the form of feedback. He draws a close analogy between brain and computer, and even between the waking and sleeping state of the brain and the condition of a computer when handling many programs as against times when it is idling, and when inputs and outputs are markedly diminished.[11]

The Peterfreund-Schwartz system model of psychoanalysis is a thorough translation of information theory into psychoanalytic terms, and vice-versa, in some ways reminiscent of Karl Deutsch's exercise in translation into the language of political science. But at no point does it appear that anything is added beyond an exercise in vocabulary; thus Peterfreund's listing of the four features of the "optimal analytic process" represents a translation of ordinary psychoanalytic terminology into the language of programming:

1. In an optimum analytic process a vast amount of information becomes available for processing, much of which may be totally new or else never before fully available. . . .

2. In an optimum analytic process information can be processed extensively. As a result, new programs can be formed, existing programs can be reprogrammed, monitoring standards can be changed, the scope of high-level control can be extended, and inconsistencies in programming eliminated. . . .

3. The optimal analytic process makes possible the full activation of many partially activated control systems related to drives and emotions. Clinically, this results in the phenomenon of abreaction. . . .

4. In the optimal analytic process the analyst plays an especially vital role. He brings new information and new programming possibilities. . . .[12]

Thus what in older language would be called a patient's "acting out" of emotional conflicts that have not been assimilated by the patient's ego are now called "subroutines which influence behavior in pathological ways but which are not brought into the 'recompute' mode accompanied by awareness." But there is nothing to indicate that anything at all is added by all this to the analyst's armory of concepts and techniques except a modish new vocabulary.

Much the same can be said of the attempt of the psychiatrists Gray, Duhl, and Rizzo to import general systems theory into psychiatry.[13] Their collection, by 28 authors, among them such stars as Silvano Arieti, Ed Auerswald, Roy Grinker, Judd Marmor, Jules Masserman, Karl Menninger, Eugene Pumpian-Mindlin, Jurgen Ruesch, and Ludwig von Bertalanffy, offers 25 essays in programmatic declarations of the importance of systems theory to psychiatry with not one indication of substantive accomplishment either on the level of clinical cure or on the level of theoretical insights and formulations other than the parading of systems vocabulary and of the piety of science.

SOCIAL CASEWORK

In virtually identical language, a program for importing systems concepts into social casework has been stated by Mary Paul Janchill.[14] Social casework is seen as based on the "person-in-situation" concept; after reviewing the history of social casework theory from the early formulations (1920s) of Mary Richmond, through the importation by Gordon Hamilton of Freudian theory into casework practice, Janchill offers general systems theory as the framework most likely to prove fruitful for the field as a "bridge needed between psychoanalytic theory and social science theory." She offers a familiar listing of system characteristics as fruitful for casework. Systems are characterized by the importation of energy, by throughputs and outputs, negative entropy, information input, feedback, coding, steady states, homeostasis, equifinality, and the like. A new vocabulary is offered:

Depression..may be interpreted in psychoanalytic terms: object loss, the incorporation of the object . . . , and punishment of the object that was originally the subject of ambivalent feelings. Another way of interpreting depression is by means of its latent functional consequences: the control it gives the patient in environmental systems, such as the family and employment. Systems analysis, then, may be a reference for a larger view of symptomatology. The notion of an output that activates another input also enriches understanding of role complementarity and role induction. Communications theory and learning theory would undoubtedly have greater applicability when used with a systems reference.[15]

General systems theory would offer further advantages, among them a less normative appraisal, more value-free appraisal, whereas the present disease model, with its reference to pathology is more judgmental; without supposing pathology, it seeks to locate the forces acting upon the individual; it will enrich the understanding of symptomatology by understanding symptoms in terms of their functions "for and across systems." Systems theory, which "has had much to offer many scientific disciplines," may also serve social work in resolving "the perennial conflict of the many and the one, with the result that the person-in-situation concept can be grounded in meaning and effective practice."

The systems approach has proven attractive to the new family therapy movement, which sees the individual as part of a network or system of cognitive and affective processes generated by his family. Treating the individual apart from this network is regarded as likely to be fruitless. The vocabulary of systems theory is adopted uncritically.[16]

SYSTEMS THEORY AS POPULAR IDEOLOGY

As systems theory migrates into more and more areas, its scope becomes wider, while its imagery becomes less precise; its links with information theory, cybernetics, and operations research become less explicit, and any set of orderly routine procedures is referred to as a "system." At this point system theory loses its links with specific disciplines and becomes a mass of imagery drawn from a vast variety of sources. I repeat here the analogy drawn by Ida Hoos:

. . . the approach, as we now encounter it, resembles the geological phenomenon known as "Roxbury pudding stone" in both history and constitution. This formation, located in a suburb of Boston, Massachusetts, resulted from glacial movement, which over the miles and the centuries dragged with it, accumulated, and then incorporated a vast heterogeny of types of rock, all set in a matrix and solidified in an agglomerate mass. Many fragments still retain their original identity and character; some have undergone metamorphosis in varying degrees. In like manner, the systems approach is a kind of mosaic, made up of bits and pieces of ideas, theories, and methodology from a number of disciplines, discernible among which are—in addition to engineering— sociology, biology, philosophy, psychology, and economics.

Each discipline has its own intrinsic and fundamental conception of system, along with its own definitions, principles, assumptions, and hypotheses. But there is a dynamic which pulls them together, makes them *gemütlich*, and provides them with a mutually supportive kinship. This consists of their orientation to and emphasis on the *totality* of the experience, entity, or phenomenon under consideration. . . .[17]

From this point on in the discussion *systems* means whatever the speaker intends it to mean, regardless of its original provenance in this mass of pudding stone.

JAY FORRESTER

Forrester's work has been an important influence in the transmission of systems thinking to the level of popular ideology,[18] both through his own writings, and through his association with the Club of Rome and its presently-famous document *The Limits to Growth*,[19] which is explicitly based on Forrester's work. ("The prototype model on which we have based our work was designed by Professor Jay W. Forrester of the Massachusetts Institute of Technology. A description of that model has been published in his book *World Dynamics.*")[20]

Forrester's use of system terminology is slightly at variance with that of von Bertalanffy, the information theorists, and the cyberneticists; but his thinking is in total agreement. Apart from minor differences in terms (Forrester uses "open" system to contrast with "feedback'" system, where Ashby or von Bertalanffy would use, correspondingly, "closed" vs. "open" system),[21] Forrester's system thinking belongs to the "simulation model" school of operations research.

For Forrester, the basic model is any system that reacts to information about the performance of the system; the model is a set of integral equations that relate the variables. The equations and the mathematics thereof are straightforward and unproblematic. Of greater interest are Forrester's assumptions about the applicability of such models to the social world. His *Urban Dynamics* attempts to apply such models to the problems of urban growth and stagnation. His primary concern in *Urban Dynamics* is with the stagnation of cities, seen in terms of the interaction of housing, industries, and population. Systems are "feedback processes having a specific and orderly structure,' ' and urban growth and stagnation are assumed to be systems of this type. The model developed is said to account for urban growth, equilibrium, and decay; by maintaining the system variables within appropriate limits, the city will be brought back from decay to an appropriate state of growth or equilibrium. Urban growth and equilibrium are considered to be a function of the relation between construction for industry housing for the employed population and available land. The urban "system' is seen as consisting of several subsystems. One is concerned with business activity (new enterprise, mature business, declining industry); another represents housing (new, aging, and housing being demolished); a third subsystem represents population and includes three categories (managerial-professional, labor, and underemployed). From these fundamental categories are developed a series of dynamic "flow rates," among them such variables as housing declines, arrival rates of underemployable labor, new enterprise starts, an "attractiveness-for-migration multiplier," a "public-expenditure multiplier," and the like. Forrester includes such refinements as a variable that delays the perception of the true attractiveness of the urban area.

In the preceding equation the perception delay has been taken as 20 years. This means that a 20-year time lag occurs between the true condition of the area and the perception of that condition by remote members of the underemployed population. This is about one generation and seems to be approximately the time required to become aware of changing conditions in our social institutions. Of course this time delay is a function of communications and would have been longer a hundred years ago than it is today with television and with more rapid transportation.[22]

Without taking any further issue with Forrester at this point, the peculiarly narrow and ahistorical naivete of his thinking can be pointed out. Population shifts 100 years ago were much more massive and rapid than today, without benefit of modern communications; the eighteenth and nineteenth centuries were the age of mass transoceanic migrations and were generated and ended by forces other than the variables included in any of Forrester's models. Ignored also is the obvious fact that the attractiveness of an urban area for "underemployed" persons (presumably this is all a euphemism for southern black poor coming north or poor white Appalachians coming to Chicago to receive public assistance) may be generated by conditions at their point of origin. Conditions may be so poor that even a "declining" urban area is more attractive than the point of origin. The underemployed may not see their arrival as contributing to the decline of the city, which in fact may be due to other causes. Thus rather arbitrary normative judgments are built into the technical apparatus, more or less disguised in technical terms.

Class interests are also built in, in the form of "managerial-professional birth rates," "managerial arrival rates," and "managerial departures.'7

Forrester's variable lists are instructive in the image they offer of the city as material to be administered in all its dimensions; the lists include new enterprise, mature business, declining industry, premium housing, worker housing, underemployed housing, managerial-professional workers, labor, the underemployed, manager/housing ratio, labor/housing ratio, manager/job ratio, underemployed/housing ratio, labor/job ration, tax ratio needed, underemployed to labor net, underemployed training program, underemployed arrivals and departures, labor arrivals, labor departures. Behind this apparatus remain arbitrary and unexamined assumptions and value judgments.

Other assumptions are stated more explicitly.

Complex systems have special responses which cause many of the failures and frustrations experienced in trying to improve their behavior. As used here the phrase "complex system" refers to a high-order, multiple-loop, nonlinear feedback structure. All
social systems belong to this class. The management structure of a corporation has all
the characteristics of a complex system. Similarly, an urban area, a national government, economic processes, and international trade all are complex systems. Complex
systems have many unexpected and little understood characteristics.[23]

It is "an interlocking structure of feedback loops." This loop structure

. . . surrounds all decisions, public or private, conscious or unconscious. The processes
of man and nature, of psychology and physics, of medicine and engineering all fall
within this structure." But while simple systems may have only a few levels (exemplified in a small number of integrating equations), a system "of greater than fourth or
fifth order begins to enter the range here defined as a complex system. *An adequate
representation of a social system, even for a very limited purpose, can be tenth to hundredth order.* The
urban system in this book is twentieth order.[24]

With all respect to Professor Forrester, I would submit that any social
system is of "infinite" order and, hence, intrinsically unamenable to treatment even by very large computers. In any event Forrester argues that complex systems must be dealt with by computer simulation methods just because
they are so difficult; complex systems are among other things, "counterintui-
tive," are "remarkably insensitive" to changes in system parameters, "stub-
bornly resist policy changes," contain "influential pressure points, often in
unexpected places, from which forces will radiate to alter system balance." In
addition, he tells us, they "counteract and compensate for externally applied
corrective efforts by reducing the corresponding internally generated action
(the corrective program is largely absorbed in replacing lost internal action)"—though how this differs from stubborn resistance to policy change is
not clear—their long-term reactions to new policies may be the reverse of
their short-term reactions; and they tend to low performance.[25]

The cure offered is simulation modeling, which will adequately describe
the "structure and interrelationships of a system." Models are not merely
essential, they are inescapable; even thoughts and images about things are
models, less valid perhaps, but models nonetheless. Model validity is judged
in terms of the purpose of the model. For models to be used to solve problems,
the model must itself be one that generates the problem; furthermore, it must
be a closed model, uninfluenced by variables external to it. Once such a
problem-generating model is created, it can be manipulated to reduce the
problem.

Forrester sees social-science problems as a failure in model building:

In the social sciences failure to understand systems is often blamed on lack of data. The barrier to progress in social systems is not lack of data. We have vastly more information than we use in an orderly and organized way. The barrier is deficiency in the existing theories of structure. The conventional forms of data-gathering will seldom produce new insights into the details of system structure. Those insights come from an intimate knowledge of the actual systems. Furthermore, the structuring of a proper system theory must be done without regard to the boundaries of conventional intellectual disciplines. One must interrelate within a single system the economic, the psychological, and the physical. When this is properly done, the resulting structure provides nooks and corners to receive fragments of our fabulous store of knowledge, experience, and observation.[26]

Yet, Forrester's own model of the city is one that occurs in a strange conceptual isolation; that is to say, his concept of "the city" does not appear to resemble any historical reality. He speaks of "The City—Master of its own Destiny" and explicitly argues that "the urban dynamics exposed by this investigation imply that the city can change from the inside. Whether or not it will . . . depends primarily on political leadership and improved public understanding leading to support for policies that can reverse urban decline. A city need not wait for the national problems of poverty and the underemployed to be solved."

"The city, by influencing the type and availability of housing, can delay an increase in the immigration rate until internal balance is reestablished." It can remove aging housing before imbalances are created on a continuous basis; it can establish industrial parks within decayed residential areas, and its regulations and tax policies can be designed to attract appropriate industries. The only hardship to result may be that of relocation, caused by three "streams of change"—slum demolition, "voluntary relocation from underemployed housing to worker housing" as economic mobility makes it possible, and the economic relocation caused by replacement of older housing by newer housing and by the replacement of declining industry by new enterprise.

The city has been presented here as a living, self-regulating system which generates its own evolution through time. It is not a victim of outside circumstances but of its own internal practice. As long as present practices continue infusion of outside money can produce only fleeting benefit, if any. If the city needs outside help, it may be legislative action to force on the city those practices that will lead to long-term revival. Such outside pressure may be necessary if internal short-term considerations make the reversal of present trends politically impossible. The revival of the city depends not on massive programs of external aid but on changed internal administration.[27]

Forrester reveals much confusion in his image of "the city." In the vast patchwork of American government forms—federal, state, county, and city governments, not to mention independent commissions, "authorities," and regulatory bureaucracies—"the city" is perhaps the weakest of all these forms. Its ability to "master its own destiny" is much less than Forrester assumes. The entire movement toward regional planning in the United States is in part a recognition that urban agglomerations extend vastly beyond the old municipal political boundaries; by now it is commonplace to note that a city like New York (and countless others) is a center of economic and social activity that reaches far beyond the limits of the five boroughs, but that the political "reach" of the city is limited to boundaries established generations ago. To assume that cities are unaffected by regional influences, by national and international corporations, by unfriendly state legislatures, as Forrester appears to assume, does not fit any urban reality available to study. All these considerations are commonplace, but they are apparently unknown to or rejected by Forrester.

Furthermore, his notion of political solutions appears to be strangely naive. If cities are not wise enough for their own good, solutions will be imposed upon them for their own good. But from where will this higher wisdom come? Are state and federal politicians and legislators significantly wiser and more statesmanlike than urban politicians? Few are likely to claim this; the unspoken source of Forrester's higher political wisdom probably lies with the bureaucratic planner, whose affinity for benevolent authoritarianism is obvious.

But even worse, Forrester's image of the city is completely out of contact with current political and economic realities. He appears to believe that the economic world is still a world of small business that cities can "attract." Major sectors of the economy, among them government and large national and international corporations, have managed to liberate themselves from the market and therefore are in a position to dictate terms to nations as well as to any city and therefore are unlikely to "respond to" attractions, but will prefer to create their own attractions. But this does not register with him. In this connection the sector of the economy that remains closely related to the free market itself—an ever-shrinking sector of the economy—is for the most part the declining small business. Such firms are increasingly marginal and unlikely to add to the economic health of the city. In addition, his notion of excluding underemployed labor from the city has an authoritarian tone.

His image is an abstraction to the point of meaninglessness. There are cities whose entire planning processes (by any name) are at the service of investing corporations, others that are exploited by surrounding suburbs, others that have no intrinsic relation to their surroundings (e.g., political capitals such as Washington D.C., which are supported entirely by tax revenues and have little or no "natural" relation to their environments in terms of industry, farming, or transportation; military bases and settlements are in a similar position).[28]

Even more, Forrester fails to make clear what things are "internal" to the city, and what things are "external." Since the old geographical boundaries of cities are meaningless with respect to their actual growth and reach, where are the social boundaries of the city when figures prominent in politics and economics (e.g., a Rockefeller) are able to act on a variety of levels—politically, on local, state, and national levels; economically, as financial forces; and on many levels through their access to or control of informal power relationships through "interlocking directorates" and influence networks? These commonplaces appear to be ignored in Forrester's scheme and find no place within his model of the urban system.

His model, though internally consistent, appears to fit no reality other than that of its own construction.

WORLD DYNAMICS

Perhaps finding the city too limited a problem, Forrester has set his sights on the world as a whole. Invited to a conference by the Club of Rome in June 1970 in Bern, Switzerland, Forrester was asked to contribute to their project on the predicament of mankind by adapting the models he had earlier developed for industrial and urban dynamics to the world as a totality. In July 1970 the group was invited to Cambridge, Massachusetts, to examine the M.I.T. system dynamics approach, and the world dynamics model based on his earlier approach was accordingly developed by Forrester.[29] Its results are embodied both in *World Dynamics* and, in greater detail, in the Club of Rome's currently famous *The Limits to Growth*.[30]

The Club of Rome deserves some passing attention. William Watts describes it:

. . . an informal international association, with a membership that has now grown to approximately seventy persons of twenty-five nationalities. None of its members holds public office, nor does the group seek to express any single indeological, political, or national point of view. All are united, however, by their overriding conviction that the major problems facing mankind are of such complexity are so interrelated that traditional institutions and policies are no longer able to cope with them, nor even to come to grips with their full content.

The members include Dr. Aurelio Peccei (described as its prime moving force), who manages a consulting firm for engineering and economic development and is affiliated with Fiat and Olivetti; Hugo Thiemann, head of the Batelle Institute in Geneva; Alexander King, scientific director of the OECD; Saburo Okita, head of the Japan Economic Research Center, Tokyo; Eduard Pestel of the Technical University of Hanover; Carroll Wilson of M.I.T.[31]

The argument of *The Limits to Growth,* though developed in great detail, is simple in outline; it is primarily a development of Malthusian assumptions. Population is outgrowing food supply, and industrial growth is threatening to exhaust natural resources within a century. In addition, pollution of all kinds is threatening the chain of life itself. The world system is driving close to or perhaps even beyond all safe limits and is threatened with collapse within a century at most. World society must therefore be transformed from one of uncontrolled growth in industrialization, consumption, and in population, to a society in equilibrium, a stable "no-growth" world system.

The formal structure of his simulation model is the same; the content now is a set of variables measuring population, resources, pollution, and "the qualtiy of life." Five "levels" of variables were chosen, under the headings: population, capital investment, natural resources, fraction of capital devoted to agriculture, and pollution. Within each level are clusters of variables, such as birth and death rates, rates of natural-resource consumption, and the like. The world problem is the same as the urban problem: how to find stability and equilibrium. (An important variable, "quality of life," is measured in terms of material standard of living, food supply, pollution, and crowding. The arbitrary definition of quality of life in these terms apparently goes unremarked.)

The conclusions of the Club of Rome report, as with Forrester's work, have by now entered the main stream of public opinion.

The work has had its opponents; the arguments have been subject to detailed analysis in a number of sources, of which three may be cited here for their caliber: John Maddox's *The Doomsday Syndrome* addresses itself primarily to claims that life is threatened by pollution and that resources are near exhaustion;[32] Peter Passell and Leonard Ross, in *The Retreat from Riches,* focus primarily on the opposition to economic growth.[33] Neither work addresses themselves primarily to the systems aspect of the population-pollution-exhaustion-of-natural-resources argument.

The most detailed and careful analysis of Forrester's work is offered by the research team at Sussex University's Science Policy Research Unit and is embodied in a direct retort to the Club of Rome, *Models of Doom.*[34] In this work, 13 scientists—among them economists, mathematical physicists, biologists, political scientists, and social psychologists—subjected the Club of Rome report to careful dissection. Their criticisms prove highly apposite to any assessment of the pretensions of systems thinkers. Among these criticisms are the following major points:

1. The Forrester model fails to include important technological and social feedback mechanisms that have in the past proven important in producing technological and social change. These mechanisms were deliberately excluded because they were considered to be too slow to prevent disaster. The models, then, are "unrealistically deterministic"[35] and slanted towards a doomsday picture.

2. The models use world average figures both for parameters and for relationships between parameters. This in turn leads to "rigid and unrealistic assumptions about the structure of distributions within the world system. Consequently it may be impossible to make sensible forecasts with such a highly aggregated model."[36]

3. Forrester's system dynamics technique is inflexible and contains approximations that lead to considerable "rounding errors," which can be large and can influence results if sensitive feedback loops are present.[37]

4. Available statistical techniques have been neglected, which makes it difficult to separate the simultaneous effects of variables such as food, wealth, crowding, pollution, birth and death rates. Where the data base is so poor and subject to such a wide range of interpretation, the effects of bias can be large.

5. The problem of extrapolation is unresolved; "during computer runs some parameters take values outside those so far experienced in the world. This means that several look-up tables have to be extrapolated well beyond the range of data available. . . . Serious errors of extrapolation are more likely to occur when relationships are assumed to be multiplicative rather than additive, especially if the errors are all biased in the same direction. Furthermore, the fact that a model appears to fit historical data within a certain range is not a guarantee of the model's validity within that range, let alone an indication that it may be used to extrapolate outside it. Many different combinations of mechanisms can give rise to the same pattern for a few major variables. Errors . . . may compensate for each other within the fitted range . . . but may not continue to do so outside it."[38]

6. Predictions of the "imminent" exhaustion of natural resources (iron, gold, copper, etc.) have been made for almost 100 years; technological innovations have continually falsified these predictions, and are continuing to do so in ways ignored by the model, which also ignores the continuing rate of new discoveries.[39]

7. A variety of assumptions built into the model are demonstrably false. These include: that low-quality ores of important materials do not exist in large quantities; that there are few geographic areas left to explore (the level of detail of exploration leaves vast room for discoveries; "Explored countries are still producing surprises").

8. The Forrester model assumes that resource depletion will be one of the

major modes of "collapse;" the main reason here is the assumption "of fixed economically available resources, and of diminishing returns in resource technology. Neither of these assumptions is historically valid. . . . Meadows looked at this sort of possibility and found that collapse still happens because available resources are fixed. We have argued that they are finite (not fixed), but that what is available in the earth's crust is (almost) infinitely greater than what is in fact assumed in World 3."[40] If only very modest increases in rates of resource discovery, recycling, and economy of use in industry occur, "then the resource mode of collapse in World 3 will be avoided and there will not be any net drain on 'available' reserves."

9. The population subsystem is highly insensitive to actual behavior.

10. The capital subsystem "assumes inflexible relationships and constants throughout which make overshoot and collapse typical modes of behavior of the model. It excludes the possibility of adaptive flexible response to changing circumstances—one of the main characteristics of real world behaviour of the economy. . . . The M.I.T. model has departed from its own declared aim of making a model of physical capital. It has done so in such a way as to make an extremely implausible assumption about life of capital, based on national accounts data for one year only.[41]

More generally, with reference to the general approaches adopted by Forrester and Meadows, the Sussex team observes:

Although Forrester deserves credit for his original formulation of the world model, World 2 emerges rather poorly from our examination . . . (it) must be considered unsatisfactory . . . because the results are so sensitive to the inclusion of small rates of technological change and resource discovery (indeed the threat of an early catastrophe from physical limitations is removed), it seems that some very important factors have been omitted from the model . . . in the conditions Forrester envisages, local catastrophes would occur long before the time predicted by World 2 for "collapse." So not only are the results of World 2 nothing like as invariant to the inclusion of "conventional responses to economic and social problems" as Forrester has claimed, they would also not be invariant to inclusion of a more realistic world structure. . . .

In the light of what has been said above, the categorical nature of some of the statements and conclusions in *The Limits to Growth* are open to very severe criticism. Statements like: "The limits to growth on this planet will be reached in the next one hundred years," "The basic behaviour mode of the world system (is the same) even if we assume any number of technological changes in the system;" "Even the most optimistic estimates of the benefits of technology in the model did not prevent the ultimate decline of population and industry or in fact did not in any case postpone the collapse beyond the year 2100;" do not appear to stand up to examination. Neither does another confident statement in the book that "we do not expect our broad conclusions to be substantially altered by further revisions."[42]

The Sussex team is well aware of the ideological background for the Club of Rome report; Harvey Simmons (in Chapter 13: "System Dynamics and Technocracy") draws a parallel between the technocracy movement of the 1930s and the approach of Forrester and Meadows:

First, there is the belief that engineering techniques can be used to indicate both the source of our problems and some possible solutions. Second, there is a shared scepticism about the ability of the citizen to understand through ordinary processes both the nature and possible resolution of these problems. Third, there is the link between scientists and a disinterested but prominent and influential elite. In the case of Technocracy, this elite was composed of businessmen; in the case of the system dynamics group, and appropriately enough in the 1970s, this elite is composed of men who are at the top of various knowledge institutions: research institutes, foundations or management consulting firms. Fourth, there is an immense, and therefore, dramatic simplicity of analysis. Fifth, since both groups share certain messianic qualities—a common faith, shared objectives within the group, and even a desire to proselytise—they can easily be viewed as movements.[43]

The prophetic nature of the movement is further suggested by Simmons' comments on the lack of empirical basis in Forrester's work.

One of the problems with Forrester's work here (*Urban Dynamics*) . . . is that he supplies virtually no evidence for the basis of his equations. Forrester admits he has been criticised for constructing a model of an urban system without any apparent acquaintance with the literature on urban systems but he excuses this by claiming that "the book comes from a different body of knowledge, from the insights of those who know the urban scene first hand, from my own reading in the public and business press, and from the literature on the dynamics of social systems for which references are given." (The literature on the dynamics of social systems consists of six references, all to publications by Jay W. Forrester. They are the only references in the entire book).[44]

Simmons points out the hidden utopian message within the message; once the pursuit of economic growth is abandoned, frantic striving will end, and men will pursue their "spiritual" side—education, art, music, religion, scientific research—all will flourish.

Like the great prophet of world salvation through world breakdown, Karl Marx, their apocalyptic visitions of the immediate future are tempered by the glittering image of utopia barely discernible through the fire and brimstone that rages in the historical foreground. This is not to denigrate the beliefs of the Forrester Meadows school in any sense; rather, it is to suggest that they too, despite the surface appearance of scientific neutrality and objectivity, bring us a message which can only be fully understood in the context of their own beliefs, values, assumptions and goals.[45]

Marie Jahoda's concluding words in the "Postscript on Social Change" appears the most appropriate epitaph:

What, then, remains of Forrester's and Meadows' efforts? Nothing, it seems to us, that can be immediately used for policy formation by decision makers; a technique, one among several—system dynamics—of promise which needs improvement; but above all a challenge to all concerned with man's future to do better.[46]

Despite the shallowness—and even falsity—of the Forrester-Meadows systems approach, or perhaps because of it, the "environmental-ecological" movement, and with it, anti-growth, have become influential popular movements, with significant influences on public opinion and through this upon legislation.[47]

Thus on many levels, from that of scholar-intellectual to scientist-king to that of popular culture, systems theory in one form or another has emerged as an intellectual "movement" and a social force. Its meaning remains to be interpreted.

NOTES

1. Wolfgang Koehler, *The Place of Values in the World of Fact* (New York: Liveright, 1938). See F. Emery, *Systems Thinking*, (Baltimore: Penguin, 1969). pp. 59–69.
2. Andras Angyal, *Foundations for a Science of Personality* (Cambridge: Harvard University Press, 1941; reprint New York: Viking, 1972). References are to the reprint.
3. *Ibid.*, pp. xi, 2, 15–19.
4. *Ibid.*, p. 247.
5. *Ibid.*, p. 256.
6. Systems theory in psychology receives no mention in works dating back to the 1930s; thus Edna Heidbreder's *Seven Psychologies* (New York: Appleton-Century, 1933, reprinted 1961), Fred S. Keller's *The Definition of Psychology* (New York: Appleton-Century, 1937; reprinted 1965) make no mention of system theory in any sense of the term. Robert Thomson's *Pelican History of Psychology* (Baltimore: Pelican, 1968), a history to 1940, similarly makes no mention of these developments and includes only a few references on information theory. Both F. H. Allport and Gordon Allport are mentioned only with respect to their earlier presystem theory work. Thus it appears that the real impact of systems theory upon psychology is a phenomenon primarily of the late 1950s and 1960s, and is still underway.
7. Gordon Allport, "The Open System in Personality Theory," *Journal of Abnormal and Social Psychology*, Vol. 61 (1956), pp. 301–311. Extracts are reprinted in Buckley, Ed., *Modern Systems Research for the Behavioral Scientist*, (Chicago: Aldine, 1968), pp. 343–350. The authoritarian potential in writers oriented toward systems thinking has been pointed out by Stanford M. Lyman in his critique of both Gordon Allport and of Talcott Parsons, in their writings on racial problems in the United States. Thus Allport's diagnosis of the American racial situation, as offered in his *The Nature of Prejudice* (Garden City, N. Y.: Anchor, 1958), Part III: "Reducing Group Tensions" contains an implicit authoritarianism in suggesting forcible interference in socialization and other institutional processes. For fuller details, see, Stanford

M. Lyman, *The Black American in Sociological Thought: A Failure of Perspective* (New York: Capricorn Books, 1972), Ch. 5, especially pp. 135–137. Since Allport's writings on prejudice, however, predate his espousal of systems thinking in his 1960 article, it is not certain that this can be attributed to the systems element in his thinking; nevertheless, the affinity is clear.

8. Emanuel Peterfreund, in collaboration with Jacob T. Schwartz, *Information, Systems, and Psychoanalysis—An Evolutionary Biological Approach to Psychoanalytic Theory* (New York: International Universities Press, 1971); Psychological Issues Monograph 25/26. Schwartz is a mathematician.

9. *Ibid.,* p. 40.

10. *Ibid.,* p. 75.

11. *Ibid.,* p. 246.

12. *Ibid.,* pp. 352–354.

13. Frederick J. Duhl, William Gray, and Nicholas D. Rizzo, editors: *General Systems Theory and Psychiatry* (Boston: Little Brown, 1969).

14. Sister Mary Paul Janchill, R.G.S., "Systems Concepts in Casework Theory and Practice," *Social Casework* (February 1969), pp. 74–82.

15. *Ibid.,* p. 80.

16. Lynn Hoffman and Lorenze Long, "A Systems Dilemma" (Introduction by Edgar H. Auerwald), *Family Process,* Vol. 8, No. 2 (September 1909), pp. 211–234.

17. Ira R. Hoos, *Systems Analysis in Public Policy: A Critique,* p. 27, (Berkeley Calif. Univ of Calif Press, 197).

18. His principal works are: *Industrial Dynamics* (Cambridge: M.I.T. Press, 1961); *Principles of Systems* (Cambridge: Wright-Allen, 1968), Second Preliminary Ed.; *Urban Dynamics* (Cambridge: M.I.T. Press, 1969); *World Dynamics* (Cambridge: Wright-Allen Press, 1971).

19. Donella H. Meadows, Dennis L. Meadows, Jørgen Randers, William W. Behrens III, *The Limits to Growth—A Report for the Club of Rome's Project on the Predicament of Mankind* (New York: Signet, 1972, reprint of Universe Books ed.).

20. *Ibid.,* p. 26.

21. Forrester, *Principles of Systems,* p. 1–5.

22. Forrester, *Urban Dynamics,* p. 25.

23. *Ibid.,* p. 107.

24. *Ibid.,* p. 107. Emphasis has been added.

25. *Ibid.,* p. 109.

26. *Ibid.,* pp. 113–114.

27. *Ibid.,* pp. 128–129.

28. In this respect, see the important discussion in James O'Connor, *The Fiscal Crisis of the State* (New York: St. Martin's Press, 1973), *passim,* but especially Ch. 3, "Political Power and Budgetary Control in the United States." See also Oliver Williams, "A Typology of Comparative Local Government," *Midwest Journal of Political Science* (May 1961), p. 160, cited by O'Connor. O'Connor underestimates the autonomy of bureaucratic structures, but otherwise marshals important data. For information on the relative lack of autonomy on the level of state and local government, especially at the city level, see, Ira Sharkansky, *The Politics of Taxing and Spending* (Indianapolis: Bobbs-Merrill, 1969), Ch. IV, "Taxing and Spending in State and Local Governments": ". . . the economic and institutional problems of state and local officials are, if anything, more difficult than those faced by federal officials. State and local governments do not have the ability to control the growth and stability of the economy. Moreover, state constitutions and laws include a number of limitations that federal policymakers avoid: rigid prohibitions or restrictions against indebtedness or increases in tax rates, requirements for balanced budgets, and earmarked revenues." (p. 83).

29. Forrester, *World Dynamics,* pp. vii–viii.

30. Donella H. Meadows, Dennis L. Meadows, Jørgen Randers, and William W. Behrens III, *The Limits to Growth—A Report for the Club of Rome's Project on the Predicament of Mankind,* with Foreward by William Watts, President of Potomac Associates (New York: Signet, 1972).

31. *Ibid.,* pp. ix–x.

32. John Maddox, *The Doomsday Syndrome* (New York: McGraw-Hill, 1972).

33. Peter Passell and Leonard Ross, *The Retreat from Riches* (New York: Viking, 1971).

34. H. S. D. Cole, Christopher Freeman, Marie Jahoda, and K. L. R. Pavitt, eds., *Models of Doom—A Critique of The Limits to Growth* (New York: Universe Books, 1973), published in Great Britain as *Thinking About the Future* (Chatto & Windus, Ltd., London). It contains a brief "Reply" by the authors of *The Limits to Growth.*

35. *Ibid.,* p. 27.

36. *Ibid.,* pp. 27–28.

37. *Ibid.,* p. 30.

38. *Ibid.,* p. 32.

39. *Ibid.,* pp. 36–41.

40. *Ibid.,* p. 41. "World 3" refers to the model constructed by Forrester for *The Limits to Growth,* more elaborate than World 2, described in *World Dynamics.*

41. *Ibid.,* pp. 65, 71.

42. *Ibid.,* pp. 132–133. The words quoted above are by H. S. D. Cole and R. C. Curnow, Ch. 9, "An Evaluation of the World Models."

43. *Ibid.,* p. 193.

44. *Ibid.,* p. 196.

45. *Ibid.,* p. 207.

46. *Ibid.,* p. 215.

47. Prominent among these are the English economist, Ezra J. Mishan and John Kenneth Galbraith. The latter's views are discussed in Passell and Ross, *The Retreat from Riches* (New York: Viking, 1971). Mishan's primary works are *The Costs of Economic Growth* (New York: Praeger, 1967) and *Technology and Growth—The Price We Pay* (New York: Praeger, 1969). The latter is a close revision of the former.

Systems Theory as Ideology

The prerequisite of a stable order in the world is a universal body of symbols and practices sustaining an elite which propagates itself by peaceful methods and wields a monopoly of coercion which it is rarely necessary to apply to the uttermost. This means that the consensus on which order is based is necessarily nonrational; the world myth must be taken for granted by most of the population. The capacity of the generality of mankind to disembarrass themselves of the dominant legends of their early years is negligible, and if we pose the problem of unifying the world we must seek for the processes by which a nonrational consensus can be most expeditiously achieved. . . . The discovery of the portentous symbol is an act of creative orientation toward an implicit total configuration.[1]

Given that systems theory has been drawn from all kinds of fields, what can we say about its characteristics?

1. First, it is an analogy, despite the denials of many systems theorists.
2. Few operations are performed with systems theory, outside of communications theory, where the philosophy was derived after the fact; the philosophy itself does not permit operations.
3. It presumes a determinacy in science, which many scientists reject.

THE LIMITATIONS OF SYSTEMS THEORY

We now reach what is perhaps the point of this study, which is that of describing the common characteristics of all systems theories that seek to transcend specific fields.

Systems Boundaries

One concept fundamental to all system theory is the notion that a system has clear boundaries. Now in a natural science, where the experiment is an important factor, the experiment can (and must) be set up to have clear boundaries. But where extrapolations are being made from the system to reality, where there are no clear-cut experimental situations, it becomes difficult to do so, but the assumption that one can arbitrarily isolate the "system" from the total reality that is infinitely complex, this very assumption of isolability is very much in doubt. (It is, of course, not in doubt in cases where artificial boundaries can be set up.) Furthermore, the system, if it is to be a system, must not only have boundaries; it must also be a closed system, sometimes closed by the very act of constructing the experiment. Dealing with a "natural reality" is always a matter of dealing with an open reality. In some areas, no doubt, closed systems can be constructed, but not in others; one cannot make a closed system just by assumption. This is a difficulty that appears to be completely evaded by systems theorists.

The Arbitrary Choice of Systems Elements

In an experimental situation only relevant factors are part of the experiment; in fact, one of the purposes of developing the experiment is to use only relevant factors and to determine their relative weights. A constructed system, which is relevant, is usually arbitrary in its reflection of nature. Thus one can be arbitrary in experiment, but one cannot call it "nature" when one has selected the factors. Thus the statement of what is a system when dealing with an open empirical system is assumed rather than proven. It is also arbitrarily assumed that the factors are interconnected and that the specific nature of the interconnections is known.

Systems Purposes

In these statements about boundaries, the closedness of systems (closed in the sense of consisting of a logically closed system of elements, assumptions, and relationships), and the interconnectedness of elements, there is a tendency to assume that the system has a purpose. To some extent natural scientists can make a teleology of nature in many senses, one being the design of an experiment to prove or disprove that nature operates in a given way.

In other areas the teleology of nature may be a linguistic fiction necessary to reconstruct an image of nature in teleological terms in order that finite minds—which tend to personalize nature—may develop some understanding. At other levels the teleology has deeper anthropomorphic significance (as in the famous tautology: survival of the fittest = survival of the survivors). In the social sciences this imputation of teleology almost always takes the form of the notion that the system *qua* system has a purpose, a purpose other than the purposes of the individual actors or the interactions of the collective actors; that is to say, systems thinkers attribute purposes to systems and not to men.

Systems and Cooperation

Finally comes the point, derived from the above questions of relevance and of boundaries, that systems theorists tend to see systems as essentially cooperative, and with almost no exceptions, social systems are seen as cooperative. This assumption means that one can work within the framework of cooperative people who cooperate with the ends of a system. And since this is done by way of implicit assumption, the contrary notion that men do not always cooperate, that they miss and misunderstand each other, and that whatever cooperation does ensue may be a result of coercion, lack of communication, misunderstanding, and error (phenomena that are often observed)—all of which one finds difficult to integrate into systems theory. Further, when such phenomena are noticed, they are taken to represent deviation from a conceptually pure system in which the values of cooperation, coherence, harmony, and the rest, are assumed to be the normal phenomenon, with the "systems" being only certain possible ways of organizing low levels of generality statements about a reality that is essentially different from the system. In other words, any empirical phenomenon will permit a wide variety of systems postulations for the same phenomenon, with no one system capable of explaining the wide range of natural variability in natural phenomena; the "system" is, like the experiment, at most one arbitrary way of organizing a small aspect of phenomena; at best it is an analogy with which one can perform no operations.

Finally, we note that systems thinkers assume that the important elements of social and natural phenomena are quantifiable. This too is the begging of an important question.

SYSTEMS THEORY AS PHILOSOPHY

Systems theory is the latest attempt to create a world myth based on the prestige of science. In earlier times such myths had to be constructed around other images possessing the power to capture men's imaginations; 500 years ago intellectuals would elaborate the images and vocabulary of theology; later, of philosophy; today, they deal in the vocabulary of science, or rather, of a debased philosophy of science.

The "philosophy" offered by systems theorists is not in any respect a unified philosophy. The specific claim by men such as Laszlo and Pepper to have provided, out of systems thinking, a "new answer to the meaning of life"[2] is pretentious nonsense. Systems theory provides a large number of such promises that are never fulfilled. The basic thought forms of systems theory remain classical positivism and behaviorism. As epistemology, it leaves philosophy no further along in resolving the Cartesian dualism; it attempts to resolve this dualism by mechanizing thought and perception, or rather by constructing mechanical models of thought and perception.

It offers nothing new to epistemology or to the problem of the Cartesian dualism. There remains no point at which one can say that here is the link between subjectivity and material processes; the solution, insofar as they can be said to have one, is to legislate subjectivity out of existence. Thus the procedure of the cyberneticians, nerve-network analysts, and communications theorists is mechanizing subjectivity—that is, mechanizing human beings—and of personalizing computers and other "systems."

The authoritarian implications are obvious to all. But even more, their very manner of posing epistemological problems indicates a failure to absorb the simplest lessons of recent philosophy, sociology, and even physics. The major unsolved problem of epistemology is how

. . . we come to know our own mental life and that of others. This is a side of epistemology which has been steadily neglected since the very earliest times. Philosophers have devoted endless trouble to discussing how we come to be aware of physical objects and how far subjective elements enter into our experience of them. They have talked as if our world consisted entirely of such objects, and as if the knowledge of them were our chief intellectual concern. Yet the most significant of our experiences lie in our relations with other people, and the nature and extent of the knowledge which we can have of other people is a question of equal importance with the first. Dilthey is the first philosopher in any country to tackle the question seriously and systematically.[3]

The growing attention paid to Dilthey's work is but one symptom of a "Ptolemaic revolution" which only now is beginning to seep into philosophy and perhaps last of all into American social science, despite the rearguard action of positivism.[4] Man is once again at the center of his universe. But aside from this influence, the now hopelessly outmoded thought patterns of the systems men is revealed within modern physics itself; not many have attempted to draw the appropriate conclusions from these developments, least of all the systems thinkers. The basic point is that the findings of "modern science" are of such a nature as to render systems conceptions outmoded before they were formulated.

. . . one of the most important features of the development and analysis of modern physics is the experience that the concepts of natural language, vaguely defined as they are, seem to be more stable in the expansion of knowledge than the precise terms of scientific language, derived as an idealization from only limited groups of phenomena. This is in fact not surprising since the concepts of natural language are formed by the immediate connection with reality; they represent reality. It is true that they are not very well defined and may therefore undergo changes in the course of the centuries: just as reality itself did, but they never lose the immediate connection with reality. On the other hand, the scientific concepts are idealizations; they are derived from experience obtained by refined experimental tools, and are precisely defined through axioms and definitions. Only through these precise definitions is it possible to connect the concepts with a mathematical scheme and to derive mathematically the infinite variety of possible phenomena in this field. But through this process of idealization and precise definition the immediate connection with reality is lost. The concepts still correspond very closely to reality in that part of nature which had been the object of the research. *But the correspondence may be lost in other parts containing other groups of phenomena.*[5]

The scientific world view, then, is at an end.[6] The notion of scientific certitude is now overthrown, not merely on the atomic level of quantum physics, not merely on the level of the replacement of old mechanical notions of certainty and causality by probabilistic concepts, but within probability, the notion that *probabilities are themselves fixed, stable, closed is illusory.* Thus the vast constructions of the cyberneticians, based as they must be on "simplifying assumptions" about probabilities that are stable, on which the entire edifice of automatons is built, is overthrown.

Equally important is the point made by Heisenberg about the loss of the correspondence of concepts to reality when extended beyond their proper range. Systems philosophy is an attempt to stretch a set of concepts into a metaphysic that extends beyond and above all substantive areas. But the extension of conceptualization into regions of empty abstraction appears futile; the futility of systems conceptualization reveals itself in every attempt at application. Concepts must be drawn from substantive areas of experience and must be saturated in concrete substantive experience. But once the concreteness of social and historical experience is confronted, systems are revealed as hopelessly rigid, petrified, and immovable, no matter how many "variables" they contain. The development of "scientific" formalisms intended to be "objective" yet comprehend the totality of experience is impossible by definition.

The road to reality is traversed through everyday language.

Heisenberg himself touches upon the biologistic origins of systems theory, only to veer away.

Concepts like life, organ, cell, function of an organ, perception have no counterpart in physics or chemistry. On the other hand, most of the progress made in biology during the past hundred years has been achieved through the application of chemistry and physics to the living organism, and the whole tendency of biology in our time is to explain biological phenomena on the basis of known physical and chemical laws. Again the question arises, whether this hope is justified or not.

Just as in the case of chemistry, one learns from simple biological experience that the living organisms display a degree of stability which general complicated structures consisting of many different types of molecules could certainly not have on the basis of the physical and chemical laws alone. Therefore, something has to be added to the laws of physics and chemistry before the biological phenomena can be completely understood.

With regard to this question two distinctly different views have frequently been discussed in the biological literature. The one view refers to Darwin's theory of evolution in its connection with modern genetics. According to this theory, the only concept which has to be added to those of physics and chemistry in order to understand life is the concept of history. The enormous time interval of roughly four thousand million years that has elapsed since the formation of the earth has given nature the possibility of trying an almost unlimited variety of structures of groups of molecules. Among these structures there have finally been some that could reduplicate themselves by using smaller groups from the surrounding matter, and such structures therefore could be created in great numbers. Accidental changes in the structures provided a still larger variety of existing structures. Different structures had to compete for the material drawn from the surrounding matter and in this way, through the "survival of the fittest," the evolution of living organisms finally took place. There can be no doubt that this theory contains a very large amount of truth, and many biologists claim that the addition of the concepts of history and evolution to the coherent set of concepts of physics and chemistry will be amply sufficient to account for all biological phenomena. One of the arguments frequently used in favor of this theory emphasizes that wherever the laws of physics and chemistry have been checked in living organisms they have always been found to be correct; there seems definitely to be no place at which some "vital force" different from the forces in physics could enter.

On the other hand, *it is just this argument that has lost much of its weight through quantum theory.* Since the concepts of physics and chemistry form a closed and coherent set, namely, that of quantum theory, it is necessary that wherever these concepts can be used to describe phenomena the law connected with the concepts must be valid too. *Therefore, wherever one treats living organisms as physicochemical systems, they must necessarily act as such.* The only question from which we can learn something about the adequacy of this first view is whether the physicochemical concepts allow a *complete* description of the organisms. Biologists, who answer this question in the negative, generally hold the second view, that has now to be explained.

This second view can perhaps be stated in the following terms: It is very difficult to see how concepts like perception, function of an organ, affection could be a part of the coherent set of the concepts of quantum theory combined with the concept of history. On the other hand, these concepts are necessary for a complete description of life, even if for the moment we exclude mankind as presenting new problems beyond biology. Therefore, it will probably be necessary for an understanding of life to go beyond quantum theory and to construct a new coherent set of concepts, to which physics and chemistry may belong as "limiting cases;" history may be an essential part of it, and concepts like perception, adaptation, affection also will belong to it. If this view is correct, the combination of Darwin's theory with physics and chemistry would not be sufficient to explain organic life, but *still it would be true that living organisms can to a large extent be considered as physiochemical systems—as machines,* as Descartes and Laplace have put it—*and would, if treated as such also react as such.* One could at the same time assume, as Bohr has suggested, that our knowledge of a cell being alive may be complementary to the complete knowledge of its molecular structure. Since a complete knowledge could possibly be achieved only by operations that destroy the life of the cell, it is logically possible that life precludes the complete determination of its underlying physiochemical structure. Even if one holds this second view one would probably recommend for biological research no other method than has been pursued in the past decades: attempting to explain as much as possible on the basis of the known physiochemical laws, and describing the behavior of organism carefully and without theoretical prejudices.

The first of these two views is more common among modern biologists than the second; but the experience available at present is certainly not sufficient to decide between the two views. The preference that is given by many biologists to the first view may be due again to the Cartesian partition, which has penetrated so deeply into the human mind during the past centuries. Since the "res cogitans" was confined to men, to the "I," the animals could have no soul, they belonged exclusively to the "res extensa." Therefore, the animals can be understood, so it is argued, on the same terms as matter in general, and the laws of physics and chemistry together with the concept of history must be sufficient to explain their behavior. It is only when the "res cogitans" is brought in that a new situation arises which will require entirely new concepts. But the Cartesian partition is a dangerous oversimplification and it is therefore quite possible that the second view is the correct one.[7]

But this question remains unsettled; at the same time, as Heisenberg indicates, we are very far from "such a coherent and closed set of concepts for the description of biological phenomena." The degree of complication in biology is too great to permit an exhaustive mathematical representation. But

if we go beyond biology and include psychology in the discussion, then there can scarcely be any doubt that the concepts of physics, chemistry, and evolution together will not be sufficient to describe the facts. On this point the existence of quantum theory has changed our attitude from what was believed in the nineteenth century. During that period some scientists were inclined to think that the psychological phenomena could ultimately be explained on the basis of the physics and chemistry of the brain. From the quantum-theoretical point of view there is no reason for such an assumption. We would, in spite of the fact that the physical events in the brain belong to the psychic phenomena, not expect that these could be sufficient to explain them. We would never doubt that the brain acts as a physico-chemical mechanism *if treated as such;* but for an understanding of psychic phenomena we would start from the fact that the human mind enters as object and subject into the scientific process of psychology.

Looking back to the different sets of concepts that have been formed in the past or may possibly be formed in the future in the attempt to find our way through the world by means of science, we see that they appear to be ordered by the increasing part played by the subjective element in the set. Classical physics can be considered as that idealization in which we speak about the world as entirely separated from ourselves . . . In . . . quantum theory, man as the subject of science is brought in through the questions that are put to nature in the a priori terms of human science. Quantum theory does not allow a completely objective description of nature. In biology it may be important for a complete understanding that the questions are asked by the species man which itself belongs to the genus of living organisms, in other words, that *we already know what life is* even before we have defined it scientifically.[8]

Equally important is the fact that quantum physics brings the return of old concepts of potentiality, of tendencies.

Probability in mathematics or in statistical mechanics means a statement about our degree of knowledge of the actual situation. In throwing dice we do not know the fine details of the motion of our hands which determine the fall of the dice and therefore we say that the probability for throwing a special number is just one in six. The probability wave of Bohr, Kramers, Slater, however meant more than that; it meant a tendency for something. It was a quantitative version of the old concept of "potential" in Aristotelian philosophy. It introduced something standing in the middle between the idea of an event and the actual event, a strange kind of physical reality just in the middle between possibility and reality.[9]

And this includes the notion that the old mechanical ideas of objectivity are no longer valid:

Let us consider an atom moving in a closed box which is divided by a wall into two equal parts. The wall may have a very small hole so that the atom can go through. Then the atom can, according to classical logic be either in the left half of the box or in the right half. There is no third possibility: "tertium non datur." In quantum theory, however, we have to admit—if we use the words "atom" and "box" at all— that there are other possibilities which are in a strange way mixtures of the two former possibilities. This is necessary for explaining the results of our experiments. We could, for instance, observe light that has been scattered by the atom. We could perform three experiments: first the atom is . . . confined to the left half of the box, and the intensity distribution of the scattered light is measured; then it is confined to the right half and again the scattered light is measured; and finally the atom can move freely in the whole box and again the intensity distribution of the scattered light is measured. If the atom would always be in either the left half or the right half of the box, the final intensity distribution should be a mixture (according to the fraction of time spent by the atom in each of the two parts) of the two former intensity distributions. But this is in general not true experimentally. The real intensity distribution is modified by the "interference of probabilities." . . . Each statement that is not identical with either of the two alternative statements—in our case with the statements: "the atom is in the left half" or "the atom is in the right half of the box"—is called complementary to these statements. For each complementary statement the question whether the atom is left or right is not decided. But the term "not decided" is by no means equivalent to the term "not known." "Not known" would mean that the atom is "really" left or right, only we do not know where it is. But "not decided" indicates a different situation, expressible only by a complementary statement.

The general logical pattern . . . corresponds precisely to the mathematical formalism of quantum theory . . . We will use the word "state" in this connection. The "states" corresponding to complementary statements are then called "coexistent states." . . . This term "coexistent" describes the situation correctly, it would in fact be difficult to call them "different states," since every state contains to some extent also the other "coexistent states." This concept of "state" would then form a first definition concerning the ontology of quantum theory. One sees at once that this use of the word "state," especially the term "coexistent state," is so different from the usual materialistic ontology that one may doubt whether one is using a convenient terminology. On the other hand, if one considers the word "state" as describing some potentiality rather than a reality . . . then the concept of "coexistent potentialities" is quite plausible, since one potentiality may overlap with other potentialities.[10]

If such considerations apply on the level of the atom, they may be expected to apply on the biological and sociological levels.

The notion of cybernetics-information-communications theory, then, of life as a set of determinate information-patterns is based on an already outmoded ontology. Ashby's idea that a black box is exhaustively known by a "complete protocol" of its past behaviors, and his use of this to construct a cybernetics is an explanation of biological and sociological matters, is so absurd as to constitute a puzzle: how was it taken seriously? The answer, of course, is the persistence of outworn thought patterns and the difficulty experienced by many scientists, let alone systems thinkers, in throwing off these habits of thought.[11]

The quicksand on which cybernetic/system theory is built—on notions of mechanical causality and determinacy—also applied to the amalgam of evolutionary theory built in, to the notion that systems evolve toward ever more complex structures, by way of purely physical and material processes. This idea has become increasingly suspect, though it continues to dominate.

In his *Autobiography* Vico points up the parellel with Epicurus, who "sets up in nature a principle falsely postulated: namely body already formed." Goethe put it another way, when . . . he criticized the "rigid way of thinking—nothing can come into being except what already is." "With the moderns," Coleridge himself wrote . . . , "nothing grows; all is made."—Growth itself is but a disguised mode of being made by the superinduction of jam data upon a jam datum. This habit of thinking permeates the whole mass of our principles. . . . "the system of . . . mechanical causality is in fact only maintainable by the surreptitious smuggling in of unreduced "immaterial influences." This is usually done by impounding them semantically in some particular word or words. Democritus, and Epicurus, postulated not only atoms, but also the famous "swerve." Today the immaterial agent of change is more likely to be impounded in some such term as "tendency" or "pattern" or "mutation" (another way of saying "change"), or "norm" or (in more up-to-date biology) "code," "message" or "information"—the whole change from e.g. a single cell to a complex living organism requiring no more than amino-acids or genes—*plus,* of course, an ability to code and decode, which last need not be unduly stressed. The trouble is, that particles *as such—* *"unproductive* particles"—cannot even arrange and rearrange themselves without more. Yet, if one credits them with immaterial "swerves" or "tendencies" and so forth, he has forgotten that those are the very things he was purporting to explain *by* them.[12]

Thus even the systems philosophers who argue against determinism still think of choices among already predetermined sets of alternatives, based upon having "sufficient" knowledge. The notion of potentiality has not penetrated.[13] I have argued that systems theory as *technique*—in its computer-based simulation models, in the mathematical foundations of cybernetics—is based on deterministic categories already undermined by modern physics. Systems theorists appear to ignore or dismiss this problem without giving it exhaustive consideration. Systems philosophers, though less comfortable about determinist thought forms, have not worked their way out of them.

But systems theory in its scientific pretensions can be examined from still another point of view. We can ask what systems theory has given us *except a new vocabulary,* that we did not have before. And, to bring the question into sharper focus, what science—new or old, theoretical or applied—can point to substantive new findings and say there are the results of systems theory? There are none to be found. Systems theorists appear to believe that concepts fruitful for a specific science—physical or social—can be developed out of systems ideas, but there is nothing to suggest that this is the case. Scientific and philosophical progress appear to have been made by a combination of thought and observation; but the thought is always substantive thought, thought related to specific problems, not thought focused either upon itself or its vocabulary. On pragmatic grounds, then, systems theory appears to make no difference.

Finally, one can query systems theory from the viewpoint of scientific method and logic, and ask if any of its propositions can be tested and thereby verified or falsified. Again, there appears to be no point at which such test can occur. A philosophy proclaiming the relatedness of all things leaves us no better off than before; on this basis, of course the most appropriate system to begin investigating is the entire universe. Lesser subsystems are there too, but there are infinitely many of them, so that investigation, as always, begins and ends arbitrarily, with arbitrarily selected problems and areas and arbitrarily selected lines of demarcation.

There seems to be no point at which the systems theorists escapes the elaboration of the system and validates it. A system of equations in astrophysics is eventually validated or invalidated by something outside the system— for example, the discovery of a new star, planet, process, which had not been known before and which the system made it possible to find. But no system is self-validating, and its validity is not established by endless refinement and elaboration of vocabulary.

Systems theory endlessly oscillates between claims as theory and claims as technique. As philosophy, it does not exist; as technique, it has limited value within tightly defined, logically closed situations. When applied to society, as a form of "social technology," all available evidence suggests that it does not work, or works disastrously. Systems theorists, as Hoos has pointed out, are then in a position to evade responsibility. If systems theory as applied social technology is criticized, they will reply that the theory was all right, but it was not carried out properly; if systems theory as philosophy is criticized, they will reply that it is to be judged as a form of praxis.

At this point it has become clear that systems theory is not a philosophy and is not a science; it is an ideology and must be considered as such.

SYSTEMS THEORY AS IDEOLOGY

The Concept of Ideology

The term *ideology* has a long history and has a variety of meanings.[14] For our purposes several are relevant. Even its popular meaning as political propaganda ("we need a better ideology to fight the enemy")[15] is appropriate to some of the more vulgarized forms of systems propaganda. But for our purposes several other meanings must be kept in mind. The earliest use of the term, Lichtheim tells us, is that of the French ideologues, especially Antoine Destutt de Tracy, for whom it means the study of the human mind and human history and culture from a naturalistic standpoint. Thereby will be gained "true knowledge of human nature, and therewith the means of defining the general laws of sociability. . . . What is 'natural' is also 'social.' Once human nature is properly understood, society will at last be able to arrange itself in a harmonious fashion. Reason is the guarantor of order and liberty."[16]

Morality, in Lichtheim's words, is seen by the ideologues as anchored in nature; the best social order corresponds to the permanent needs of man.

The antecedents of this view are, for Lichtheim, traceable to Bacon and

Descartes; the Baconian criticism of the "idols," and Descartes' "systematic overthrow" of all his opinions, in order to allow the survival of only those that pass the test of rational examination, are forerunners of the Enlightenment's application of critical reason against "unreasoning prejudices." Helvetius, described by Lichtheim as a favorite of both Marx and Nietzsche, anticipates the sociology of knowledge. ("Our ideas are the necessary consequence of the societies in which we live.")[17]

The prejudices of the mind are "the necessary fruit of social constraint and selfish interest," but "can be discredited by reason and removed by education." Lichtheim makes the important point that the French *ideologues* were the major forerunners of positivism. "For all the inherent skepticism with respect to shared beliefs, the power of rational thought was not seriously called in question. Almost a century later, Comte's positivism, notwithstanding its authoritarian features, was still rooted in the same confidence."[18]

The rationalist-positivist streak is easy to trace in systems theory, especially in Pepper's and Laszlo's attempts to derive social ethics from a mystique of systems evolution, attempts that never go beyond the programmatic stage, as Comte himself did not. One can only suppose that the attempts of the systems men along these lines can be attributed to the fate of repeating a history of which they were ignorant.

Another meaning of the term *ideology* contains a greater resonance than its earliest sense; this meaning emerges out of the Hegelian-Marxian view, a view recognized as helping to undermine the rationalist faith in reason. Hegel's notion of history as having purposes hidden to men led to the notion that the purposes men thought they were following were not the true significance of their actions and directly to the Marxian notion of "false consciousness." It was not the consciousness of men that determines their existence, but their (social) existence that determines their consciousness. Thus the alienated consciousness of men masked reality. From this point it was no great step to move to the concept of ideology as a form of consciousness that masks the play of interests.

This view can be developed to the point of extreme relativism: All ideas, forms, consciousness, and even knowledge itself are reflections of social realities. These forms are not to be dealt with "on their own terms," but are to be related to the social matrix that generated them.

It might seem that on the materialist assumptions Marx accepted as part of his conversion to socialism in 1844–1845, he was bound to arrive at a radical historicism and relativism. But although in many places the language of the *Holy Family* and the *German Ideology* (not to mention the *Communist Manifesto*) seems to support this conclusion, he did not in fact do so. He took over from his French predecessors the critical demolition of traditional metaphysics, yet he also went on ascribing a rational content to history. The rationality was a hidden one and had to be discerned in the logic of the "material" process itself, not in the "ideological" reflex it left in the minds of the participants. . . .

Marx preserved the original motive of this thinking . . . by refusing to recognize the dilemma inherent in the principle that modes of thought are to be understood as "expressions" of changing social circumstances. He took it for granted that, though consciousness is conditioned by existence, it can also rise above existence and become a means of transcending the alienation which sets the historical process in motion. The *truth* about man is one and the same for all stages of history, even though every stage produces its *illusions*. . . .

The unity of mankind, and the universality of truth, were as real to him as they were to Hegel, and it was left to his disciples to destroy the coherence of this thought by abandoning its unspoken assumptions and transforming his doctrine into a variant of positivism.[19]

Accordingly, the critique of ideology became a form of relativism with Dilthey and Weber.

History and sociology combine to make it appear that consciousness cannot transcend its time horizon, since the concept imposed upon the raw material of experience are themselves historical. Something like this had been suggested by Hegel, and following him by Marx, but they were saved from relativism by the belief that both the nature of man and the logic of history could still be grasped by an act of intellectual intuition.[20]

Systems philosophy, it should be noted, shows some affinity with the Marxist and Hegelian aspiration to grasp, by an act of intuition, both the nature of man and the logic of history, the former by algorithmic, rather than philosophical or economic methods, but it remains only an aspiration.

Lichtheim traces the evolution of the concept of ideology through more ramifications than are immediately relevant to our purpose; it is the formulation of Karl Mannheim that is appropriate here. Mannheim's *Ideology and Utopia* (1929) restates the entire problem of ideology in terms of the sociology of knowledge, especially in terms of the role played by one particular stratum of society, its intellectuals, in the creation and transmission of forms of consciousness.[21] Assuming that Mannheim's views are known well enough to require no detailed exposition, only three points are stressed here—his concepts of *ideology, utopia,* and the function of the intelligentsia. In Mannheim's words:

The discovery of the social-situational roots of thought at first . . . took the form of unmasking. In addition to the gradual dissolution of the unitary objective world-view, which to the simple man in the street took the form of a plurality of divergent conceptions of the world, and to the intellectuals presented itself as the irreconcilable plurality of thought-styles, there entered into the public mind the tendency to unmask the unconscious situational motivations of group thinking. This final intensification of the intellectual crisis can be characterized by two slogan-like concepts "ideology and utopia" which because of their symbolic significance have been chosen as the title for this book.

The concept "ideology" reflects the one discovery which emerged from political conflict, namely that ruling groups can in their thinking become so intensively interest-bound to a situation that they are simply no longer able to see certain facts which would undermine their sense of domination. There is implicit in the word "ideology" the insight that in certain situations the collective unconscious of certain groups obscures the real condition of society both to itself and to others and thereby stabilized it.[22]

Ideology, then, consists both of that form of consciousness and also of that way of interpreting and understanding the world, which justifies or maintains specific relations of power. It is, in a sense, a claim to social advantages in disguised form. *Utopia* is very much the same.

The concept of *utopian* thinking reflects the opposite discovery of the political struggle, namely that certain oppressed groups are intellectually so strongly interested in the destruction and transformation of a given condition of society that they unwittingly see only those elements in the situation which tend to negate it. Their thinking is incapable of correctly diagnosing an existing condition of society. They are not at all concerned with what really exists; rather in their thinking they already seek to change the situation that exists. Their thought is never a diagnosis of the situation; it can be used only as a direction for action. In the utopian mentality, the collective unconscious, guided by wishful representation and the will to action, hides certain aspects of reality. It turns its back on everything which would shake its belief or paralyze its desire to change things.[23]

Having reached the stage where this reciprocal unmasking has become the common property of all groups, we arrive at the fragmentation of intellectual life into a multiplicity of worlds and of styles of thought. Such being the case, what synthesis and resolution are possible, and from which segment of society can they emerge?

First, Mannheim suggests, a true synthesis would not be "an arithmetic average" of all the diverse aspirations of the existing groups of the society. "If it were such, it would tend merely to stabilize the *status quo* to the advantage of those who have just acceded to power and who wish to protect their gains from the attacks of the 'right' as well as the 'left.'" On the contrary, a valid synthesis must

. . . be based on a political position which will constitute a progressive development in the sense that it will retain and utilize much of the accumulated cultural acquisitions and social energies of the previous epoch. At the same time the new order must permeate the broadest ranges of social life, must take natural root in society in order to bring its transforming power into play. This position calls for a peculiar alertness towards the historical reality of the present. The spatial "here" and the temporal "now" in every situation must be considered in the historical and social sense and must always be kept in mind in order to determine from case to case what is no longer necessary and what is not yet possible.

Such an experimental outlook, unceasingly sensitive to the dynamic nature of society and to its wholeness, is not likely to be developed by a class occupying a middle position but only by a relatively classless stratum which is not too firmly situated in the social order. The study of history with reference to this question will yield a rather pregnant suggestion.

This unanchored, *relatively* classless stratum is, to use Alfred Weber's terminology, the "socially unattached intelligentsia" (freischwebende Intelligenz). . . . A sociology which is oriented only with reference to social-economic classes will never adequately understand this phenomenon.[24]

The intellectuals are too differentiated, too diverse in origins and occupation to be regarded as a single class, but there is a unifying bond: education, across differences of birth, status, and profession. But Mannheim sees the very uprootedness of the intellectuals, their marginality, as freeing them from the partiality of view forced upon those bound to specific social classes and occupations. Their uprootedness also prevents their assimilation to other specific classes. Their impact has already been considerable; they have infused intellectual demands into politics and, further, have "become aware of their own social position and the mission implicit in it."

One of the basic tendencies in the contemporary world is the gradual awakening of class-consciousness in all classes. If this is so, it follows that even the intellectuals will arrive at a consciousness—though not a class-consciousness—of their own general social position and the problems and opportunities it involves. This attempt to comprehend the sociological phenomenon of the intellectuals, and the attempt, on the basis of this, to take an attitude towards politics have traditions of their own quite as much as has the tendency to become assimilated into other parties.

We are not here concerned with examining the possibilities of a politics exclusively suited to intellectuals. Such an examination would probably show that the intellectuals in the present period could not become independently politically active. In an epoch like our own, where class interests and positions are becoming more sharply defined and derive their force and direction from mass action, political conduct which seeks other means of support would scarcely be possible. This does not imply, however,

that their particular position prevents them for achieving things which are of indispensable significance for the whole social process. Most important among these would
be the discovery of the position from which a total perspective would be possible. Thus
they might play the part of watchmen in what otherwise would be a pitch-black night.
It is questionable whether it is desirable to throw overboard all of the opportunities
which arise out of their peculiar situation.[25]

Several considerations emerge from these three main points in Mannheim's
argument. First, since Mannheim, intellectuals have emerged in far greater
numbers than even in his time and have come to occupy more and more
positions in the political and administrative spheres. It is difficult to say that
the intellectuals are a new class, though they are certainly a distinctive group.
A full sociology of the intellectuals remains to be developed, partly because
they are so newly arrived on the scene as a major force, partly because their
character and social base have not yet fully crystallized. But Mannheim
would surely not have excluded the possibility that the intellectuals might
emerge as something of a class, or even an estate, depending on the (then)
future direction of social development. Since we stand on the threshold of a
societal evolution in which the intellectuals might emerge, and perhaps have
already emerged, as a specific class, they might well develop an ideology
appropriate to their class interests. Further, it is equally clear that systems
theory is an attempt—another in a long line of such attempts—to formulate
such an ideology. The societal conditions favoring its emergence are clear:
the growth of administrative and technical occupational groups, now out-
numbering occupations of primary production; a society whose traditions
include an optimistic faith in "progress through scientific rationality" and an
equally naive faith in the expert; the fading of older political traditions; the
bureaucratization of more and more sectors of society; the disappearance of
intellectual traditions other than those of technology and applied science; all
of these provide fertile ground for systems theory as ideology of the administrative intellectual.

A second point that emerges is that this intellectual movement may possess
features belonging simultaneously to both "ideology" and to "utopia." Systems theory as mystification contains enough obscurantist elements to permit
both support of the status quo, or at least of major structural elements thereof,
and to permit the absorption (through "co-optation") of new dissident elements as these emerge, all at the same time. After all, administrative organizations can be systems that maintain themselves, "meet challenges," and at
the same time "evolve" by developing new "adaptive responses," as the occasion requires. Systems theory, then, provides rhetoric for new forms of opportunism. That systems theorists have displayed this sort of opportunism is not
difficult to show, both in their roles as part of the military, in counter-insur-

gency, and in the intensifying development of administrative centralization, and even more, in the sudden awakening to the social malaise of the 1960s, and with this, a demand for "relevance" for the solving of social ills. The ideology of systems theory could be said to consist of having no ideology, in the popular sense of a specific political commitment. Like the dialectic, it provides a vocabulary that permits its practitioners to celebrate and serve whatever social developments emerge over the horizon. It can be both conservative and revolutionary at the same time, perhaps in the same sense in which Marxian dialecticians are conservative with respect to their own societies and revolutionary with respect to others.[26]

We now reach the point at which we can examine the systems movement as a sociological phenomenon. We have explored its disciplinary origins, its convergence into a "movement" offering a social philosophy and claims to social significance, and its migration into a variety of administrative and political spheres. Systems theory as science, as scientific philosophy, as social philosophy, and as applied technique has been examined and found inadequate. Its authoritarian potential seems striking to all but the systems theorists. Everything that has preceded this has been, in a sense, only preparation for an understanding of systems theory in sociological terms.

THE INTELLECTUAL BACKGROUND OF SYSTEMS THEORY

Systems theory as social doctrine may be regarded as a new variant of organic or "organismic" approaches to society. This tradition has a long history and has been traced back to Hindu religious thought in the East and at least as far back as the medieval period in the West.[27]

The image of society as organism appears attractive to intellectuals, who will see themselves as the brain and nerve center of the organism, dealing as they do with symbolic and conceptual matters. Recent scientific work appears to offer hope of reviving this image. But regardless of the details of the analogy and of how the details of the new biologism differs from earlier versions, the social import is always the same in the sense that it tends toward a doctrine of increasing unification and centralization of social functions, with organizations reified and claiming a monopoly of the processes of thought, will, and action, and with human individuals reduced to the role of "cells" in the organism with their functions and sphere of action delimited from outside and from above. The question, of course, is not merely whether this systems-organic image is "true," but rather: (1) what are its probable social consequences, and (2) what can be done with the image. With respect to the first, the answer has been evident from the beginning: Systems theory appears to be the "natural" ideology of bureaucratic planners and centralizers and both expresses and fosters developments along those lines. In terms of the second question, there is no doubt that systems theory may often be useful and provocative to treat problems in a variety of spheres in systems terms, but again, the criterion will remain a pragmatic one: To what new insights and substantive results does it lead? So far there appear to be none. Although the system image of the world may be true, other nonsystemic images may prove equally true. Thus it could be argued that society seen as a totality possesses no intrinsic unity and is a mere aggregate of countless small "communities" continually emerging and dispersing with the interactions of concrete individuals, and that any image of organic "unity" is a myth that remains pure image and is seated nowhere in reality. Such a view may be of equal descriptive value and may prove equally or more fruitful in dealing with substantive problems.

To be sure, the "truth" or the value of a theory often has little relation to its currency, and so still another question emerges, the problem of why the organismic image of society reemerges in systems guise at this time. This is a difficult problem in the sociology and history of ideas, but this emergence appears to be conditioned by the bankruptcy and decline of images and philosophies of society that until now have served as unifying ideas. Societies have been conceived of as unified on the basis of the social contract, of a divinely anointed king, of a covenant, or by the right of a revolutionary elite to realize supreme values. One of the features of the present era is the exhaustion of most such unifying images; European history of the past 100 years could be described in terms of a civil war among the exponents of major conceptions of society. With the bankruptcy of notions of the social contract and the military defeat of both Nazism and Fascism, the collapse of liberalism from within, and the evident loss of faith among socialists themselves in both the theory and practice of socialism in all its variants, we may be said to live in an "interregnum."[28] In this sense there is some small core of truth in the otherwise questionable notion of the "end of ideology."

It is at such a time that the notion of society run by impartial benevolent technicians operating on the basis of actuarial logic and impersonal algorithmic methods could come to the fore as a new ideology (or nonideology), and in fact one that appears attractive to all advanced industrial societies regardless of their official eighteenth- or nineteenth-century creeds.

Related to this historical dimension is the question of the peculiarly American provenance of systems theory; though it is rapidly spreading to Europe and to the Soviet societies it has taken root most strongly in American soil. This leads to a consideration both of traditional long-lasting American values and to the specific transformations undergone by American society in recent years. The values descried by de Tocqueville, the belief in progress, the interest in applied science, the belief in education and knowledge, the taste for general ideas, would all constitute receptive soil for the rhetoric of systems theory and influence its style of presentation.[29]

Here it becomes necessary to consider briefly the recent structural changes in American society that have helped generate the vogue of systems theory and to prepare for its extension into broader fields.

The New Form of American Society

American society has undergone great changes since 1945; the extent and significance of these changes have begun to be explored in recent years by a number of observers. Among the most significant recent studies are: John Kenneth Galbraith's *The New Industrial State*, Joseph Bensman's and Arthur Vidich's *The New American Society*, David T. Bazelon's *Power in America—the Politics of the New Class*, Benjamin Kleinberg's *American Society in the Postindustrial Age*, Theodore Lowi's *The End of Liberalism*, and James O'Connor's *The Fiscal Crisis of the State*. With varying emphases and shades of ideological opinion, all offer a portrait of this society's emergent features; this portrait has not yet registered on the public mind, though it may be expected to over the next several years.

The main features of this society include the following:

1. The growth of government from a relatively insignificant factor in the economy to a major factor.

2. The rise of new forms of industry based upon technology and automation, generating a wide range of new materials and products and forcing older industries and materials into obsolescence.

3. The concentration of economic power in giant corporate enterprises that extend across the globe in their reach and dwarf what remains of small business.

4. The liberation of corporate enterprise from what remains of the "free market" partly through retained earnings, partly through government contracts, and partly through the vertical integration of manufacturing and the flow of materials.

5. Enormous changes in the American occupational structure, with more Americans engaged in administrative, planning, bureaucratic, and technical work than are engaged in primary productions.

6. The rise of a new white-collar middle class, college educated, and oriented primarily to administration.

7. The rise of what might be called the social-service state, with the state assuming more and more responsibility for health, education, and welfare, in all their ramifications; with this, the rise and spread of both governmental and private bureaucracy, replacing all earlier forms of administration, and

even more, the spread of bureaucratic style, psychology, and values, until they permeate the entire society.

8. The growth in the importance of education, or at least of diplomas, as indispensable means of entry into the job market; with this, not only the increased prestige of educational institutions and educators, but more and more years spent attending schools by more and more people, and accordingly, an expansion in the numbers of the white-collar "salariat." The government and social -services sector of the economy appears to have grown at a much faster rate than have all other sectors.

9. A shift in power and prestige away from property owners and inheritors toward the new middle classes and especially toward that segment called by Galbraith the "technostructure," a shift sometimes referred to by Burnham's term, "the managerial revolution." Though the extent and even the existence of this shift in power has been disputed by some observers, especially Marxists, there is no doubt that this new technocratic middle class has made and continues to make demands for an increasing share of power. Marxian observers appear for the most part to minimize or to dismiss bureaucracy and bureaucratic functionaries as mere instruments of power relations and to underestimate the increasing autonomy of bureaucratic structures, the increasing grasp of power attained by upper-level bureaucratic elites, and to ignore the bureaucratic mentality.

10. The two principal items of the government budget are military expenditures and social services.

11. The devolution of power away from the traditional centers of democratic society (i.e., representative institutions) toward more or less invisible elite groups who arrive at social consensus through interest-group bargaining. Nieburg refers to this as "the contract state,"[30] though he refers primarily to the close alliance between defense firms and specific government agencies whose cooperation has at times by-passed public policy.

Theodore Lowi, in his important *The End of Liberalism,* describes this process or condition as "interest-group liberalism."[31] Kleinberg describes it as "elite-pluralist politics," in which the planning and decision process has been captured by favored elites.[32]

Bensman and Vidich, in *The New American Society,* speak in terms of a network of informal power cliques, united on the basis of their perception of a common interest, who work informally to coordinate what appear to be (nominally) separate or even rival power centers and who seek to manufacture and define public opinion once these elite groups have formulated social policy.[33]

All are agreed that the new development represents a unique problem. In Kleinberg's words:

. . . this is a pluralism with a pyramidal structure, in which a new upper level of interest bargaining between favored elites sits astride the earlier system of bargaining among a variety of local and national interests, and rapidly supersedes it. What we are describing here then, is in a rather special and formal sense "pluralist politics," but it is an elite-pluralist politics of a very different sort than that which has been associated with the concept of pluralism over most of its history in the West, and certainly cannot be equated with democratic practices involving a conception of popular government or popular sovereignty in any very significant way.[34]

Given these changes in the structure, composition, and tone of American society, certain factors converge that help bring systems theory to a head as an intellectual movement.

The operations researchers, intoxicated by their partial and exaggerated successes in the development of weapons systems and in the space program, have looked eagerly at the other major beneficiary of federal budgeting, the social service sector, and have sought to transfer their technology to this sector, including crime control, urban decay, education, mass transit, health services, and the like. Within this structure of interests, the crudely self-serving nature of much of the systems literature, especially that generated by operations researchers in their headier, more utopian moods, is obvious and barely constitutes a "problem" for sociological analysis.

The bureaucrats in the social services sector, aware of the early successes and the (at one time) waxing prestige of the military systems men, hastened to adopt and adapt systems theory as bureaucratic public relations for the mystification of both their government masters (legislatures and executives) and of the public at large. Systems methods have served them not only as public relations devices, but as centralizing power instruments of the first order. To paraphrase Dahl and Lindblom, he who controls the computer counts his preferences a thousand times as against preferences of others.[35]

Both these developments have occurred simultaneously, an illustration perhaps of the elective affinities that are said to operate between ideal and material interests.

The academic world in the 1950s and 1960s, much expanded by the growing demand for education on the part of the population and by the demand for research and consulting services by both government and private organizations, developed systems ideologies of its own, echoing and resonating in response to the somewhat more "practical" systems theories of the cyberneticians and the operations researchers. These systems theories, no doubt reflecting their academic environments, were more "scholarly" in character, carried greater reference to philosophical and theoretical-scientific traditions, but openly espoused the systems theories of the engineers. The engineers, the cyberneticians, and the computer simulators have at least attempted to put their systems to work, although these methods, when applied to social services and even to military affairs, have proven extravagant failures, and although their advocates appear unable to learn from these experiences and persist in the belief that what is needed is more and better computers, bigger similations models, more elaborate programs, and the like, and although they do not advertise their failures, their systems have at least been put to the test. But the systems philosophers of science and their epigones in psychology, psychiatry, social work, and family therapy have not been put to any empirical test; indeed, there seems to be no test conceivable to which they could be put, since they dwell on the plane of vocabulary, definition, and redefinition, and they never leave it.

We appear to have, then, three sources of systems theory as ideology: (1) the engineers, the cyberneticians, and the operations researchers, located for the most part in the aerospace industry and in the sectors of the academic world that are oriented toward technology; (2) academicians, including economists, political scientists, sociologists and biologists, oriented primarily toward science or "scientism" in one form or another; (3) the bureaucrats of the social service world.

We have examined something of the views and character of the first and second groups, but we have said relatively little about the third, the bureaucrats oriented toward social services.

The Rise of the Welfare State

The welfare state is a mutation of the liberal state of the nineteenth century. While accepting the vestiges of the laissez-faire economy, the societies that comprise the welfare state—England, much of western Europe, and increasingly, the post-New Deal United States—have accepted the notion that the state apparatus must be charged with the responsibility for protecting its citizens from a variety of misfortunes—hence, the rise of pension and social security schemes, health insurance, workmen's compensation, and more recently a wide range of antipoverty programs. Hence, a new social constellation has arisen that appears to be unique in that it was unanticipated in the old liberal state and shows features distinctive from the bureaucratic machinery of the socialist countries. This constellation claims, among other things, to be nonpartisan, or nonideological. It is associated with no specific political party. It claims to have no vested interests of its own; its only bias is said to be "in favor of democracy" and concern for the poor. The growth of this bureaucracy in size and power is now clear, but a full history and sociology of the welfare state has not yet been developed. A new social group seems to have emerged—one hesitates to call it a "class," perhaps "estate" or "caste" might be equally appropriate terms—whose only historical counterpart appears to have been the mandarinate of imperial China. This new caste appears to exhibit new forms of authoritarian mentality—instrumental, service-oriented, and claiming to be nonpartisan.

Fritz Morstein Marx, in his *The Administrative State*, made early mention of the authoritarian possibilities of the new service bureaucrats.

"Bureaucracy" is an ambiguous term. . . . It might mean, *first*, the type of organization used by modern government for the conduct of its various specialized functions, embodied in the administrative system and personified more specifically by the civil service . . . *second*, a mechanistic and formal approach in carrying out such functions, literal and "inhuman" to the point of indifference toward the effects achieved. . . . *Third*, bureaucracy might mean one or two other things or both at the same time. On the one hand, it might mean the kind of government that shoulders a large burden of responsibilities in support of the economic and social order. On the other hand, it might mean a political condition in which the executive branch plays a role increasingly more important in relation to the role of the legislative and judicial branches. This is decried, we recall, as the "welfare state," but it is also spoken of sympathetically as the "service state." And, *fourth*, bureaucracy might mean government subject to control, not by the electorate, but rather by a group of power-hungry, visionary functionaries. Each of these meanings calls for closer examination.[36]

He asked, at another point, whether we were not "on the road to a new absolutism more ruthless than the old, which was often moderated by the philosophical currents of enlightenment? . . . Can we be sure that nations or leaders can maintain control over the growth of government?"[37] Finally, he attempted a typology of bureaucracies, listing among the types: (1) the guardian bureaucracy, dedicated to the common good, of which the British civil service was one prototype; (2) the caste bureaucracy, exemplified by the German civil service of the late nineteenth century; (3) the patronage bureaucracy, on the Jacksonian "spoils" model; (4) the merit bureaucracy, based on examinations and diplomas.[38] Morstein Marx indicates some difficulties with his typology, specifically that it is often difficult to classify specific examples and that the line between a merit and caste bureaucracy is often hard to draw.

But, although Morstein Marx raised the question of the antidemocratic elitist mentality and practice of the new welfare-state bureaucrat at several points in his study, he did not give the question any systematic treatment or development, other than several passing expressions of concern. Nevertheless, he saw the issue clearly enough.[39]

At several points in his writings, Ralf Dahrendorf expresses similar awareness: "The ruling political class of post-capitalist society consists of the administrative staff of the state, the government elites at its head, and those interested parties which are presented by the governmental elite."[40] Also:

Plan rationality clashes with social reality in a special way. As a claim to certainty, plan rationality always implies an absolute claim to power. But reality resists such claims. Resistance and opposition do not disappear just because the ruling groups want to see them disappear; the plan that does not tolerate contradiction does not for this reason remain uncontradicted. Thus, any plan-rational behavior must conclude many a compromise with reality.[41]

The issue has also emerged in a striking but fragmentary way in the writings of Daniel Patrick Moynihan, in his several recountings of his experience in the Johnson and Nixon administrations, and in related publications. Thus Lee Rainwater and William L. Yancey were able to document the use of propaganda techniques, the suppression of materials, and the falsification of issues by government bureaucracies and their intellectuals.[42]

In his *Maximum Feasible Misunderstanding,* Moynihan espouses the views of Elliott A. Krause.

Although . . . nominally impartial . . . public bureaucracies in fact enjoy considerable freedom to "propose public policies and to promote their adoption." . . . public bureaucracies frequently interpret the public interest as requiring them to favor the interests of one group in the public at the expense of others. With William B. Boyer he notes that "the laws which created bureaucracies and their goals can be widely and variously interpreted by the bureaucracies themselves, once they are in existence." This trait began to produce consequences only when bureaucracies began to be created for the purpose of intervening in social situations.[43]

The Moynihan-Krause view is that bureaucracies are real power factions and are able to some extent to evade the control of elected officials; they move toward "increasing justification of their actions directly with the public." They create ideologies to legitimize "bureaucratic interventions that might otherwise be resisted, if not rejected." The bureaucracy seeks "to substitute its own judgment for that of Congress and the President—and, by extension, for that of the Cabinet and the Sub-cabinet.[44]

The same theme emerges in Moynihan's history of the fate of the Family Assistance Plan under the Nixon administration. For Moynihan, a reasonable and probably viable plan was sabotaged by a variety of interest groups, among them the social work profession and their allies in various government bureaucracies, that sought to bypass government authority and to manufacture public opinion against the proposal.[45] In effect, the power of decision was removed from the government and captured by interest groups partially anchored in government bureaucracies. This important theme is a consistent counterpoint in Moynihan's historical narrative but does not receive systematic treatment there. Nevertheless, the issue exists and calls for thorough exploration in the near future.

The Elite in the Welfare State

A full-scale, systematic treatment of this theme, though one not specifically oriented to the new forms of American bureaucracy, is offered by the Dutch sociologist, Piet Thoenes, in his remarkable study, *The Elite in the Welfare State*.[46] He defines an elite as a "relatively small, adventitiously organized group, which legitimately or not, exercises authority, lays claim to exercise it, or believes it should exercise authority over the other groups with which it maintains a relationship, usually of a political or cultural nature."[47]

Such an elite creates tension between the welfare state and Democracy. Such an elite is convinced of its superiority, which is not conferred upon it from below, by those administered. Their scientific insight gives them new notions of the social contract based upon general will.

If . . . there was ever a spectre haunting Europe, it was the ghost of this general will. In Hegel it appeared as the Idea, in Marx as Ideology, in Mosca as the Formula, in Mannheim as the Plan for Freedom . . . Its theoretical bearers have been the people, the ruling class, the elite, the proletarian vanguard, the socially uncommitted intellectuals.[48]

For this group, charisma is focused not on a single person, but among the chosen few, which claims for itself a spiritual superiority and sets itself the task of changing or preserving a specific social order. It must attain power for this purpose. Further, the task imposed upon the elite by its superiority includes the right and duty of determining the destiny of others. The superiority of this elite takes the form of superior intelligence:

The basic idea of "The Intelligentsia to the Top!"—even if it may have failed in practice—has hardly ever been opposed in theory.

Intelligence became expressed by means of an intelligence quotient. No matter how furious the argument may have been over the question paper which was supposed to measure I.Q., it remains in fact no more than a technical argument; a few more refinements, so to speak, and the ideal question-paper will be discovered. As long as there isn't one, all army tests and all vocational guidance bureaus will have to put up with the second best.[49]

The elite that makes its claim to superiority on the grounds of science, is especially advantaged today; they claim the possession not only of power but of intellect. Further, Thoenes suggests, elites emerge in three stages: "the announcement of the message, the carrying out of the plan set forth in the message, and the conservation of the order achieved by carrying out the plan." In its early stages, it pronounces other elites and the order which sustains them as unjust, out of date, or immoral.

The character of the elite is determined by the way in which the task is set. This is something more than just the special Christmas gift list of this or that section of society. The plan is bound up with the common will; in other words it is the concern of society as a whole. It is certainly not (so they claim) merely the dictatorial imposition of the will of a minority. The plan means the realization of the destiny of the whole of that society which is at present in existence. . . . [But] the historical process does not simply operate automatically. There exists an adept section of society which a certain group has recognized and selected on account of its special capabilities, to come forward as midwife in a critical situation.[50]

Such groups must for a time tolerate those who have not yet seen the light or are incapable of seeing it. Once it attains power, it can move to the fulfilling of its plan. For Mannheim, the planned democratic society will bring the intellectual elite to its rightful place.

In his Chapter 5, "The Structure and Influence of Elites," Thoenes offers a typology of elite structures; three ideal types from artistic circles and three from politics. (1) *The circle,* the height of informality and offers no fixed abode. "Its members rarely gather; they often meet. . . . There are central figures and marginal figures. There is a recognized seniority; there are unwritten rules which must not be broken (for one's own sake)."[51] (2) *The salon,* located in a home "of good standing," fairly exclusive, often belonging to a wealthy patron of one sort or another, usually "sympathetic and liberal." (3) *The academy.* (4) *The conspirators,* outcasts in the darkness who are brave and united in the face of external threats of death or imprisonment. (5) *The party,* more prosperous, more established, with discipline accepted by all, and a growing place in the world. (6) *The department,* governed by paperwork, by documents marked "confidential," serves ministers good and bad and the interests of their country; one's career advances steadily, as one has mastered the administrative machinery; in the world of private business, there are no such things as permanent appointments, but the civil service cannot do without them.

All such elite groups have their circles of influence and their response-groups—some larger, some smaller. The elite supplies inspiration; the response groups supplies admiration, especially important for newly emerging elites. But once the elite establishes itself, especially in the welfare state, its character changes.

Thoenes explicitly defines the welfare state as a new form of society. He argues that early founders of this society, such as William Beveridge, and current students of it, such as Richard Titmuss, try to "steer a safe course between the rocks of liberalism and socialism," and apparently for political reasons say nothing to reveal its essentially new character.

The realization that the Welfare State is a new form of society has begun to dawn but it has not penetrated far as yet, partly because the characterization of a society is always a more or less precarious business. Each specific society undergoes its own special type of evolutionary process. It includes a number of slowly decaying elements left over from earlier periods, and a number of emergent elements which belong to the future. This is often an element of arbitrariness in the lines of demarcation of what may be called the characteristics of the period described. In this present treatment the Welfare State will be described by means of contrasts with its predecessor, the liberal, and its rival, the socialist, societies. It can also be distinguished from the fascist and communist forms of social welfare, not only on account of its democratic aspects, but also by the smaller stress which it puts on collectivity.[52]

There is no "pure" welfare state in existence, and the Netherlands, Sweden, Britain, and the like each bear distinctive features. The welfare state inherits the capitalist system of production from its previous liberal period; but the system has undergone modifications, partly through its own internal development, partly under the influence of the welfare state. But two fundamental elements are retained—private property, and the profit motive. Its distinctive feature, of course, is that the community, acting through the state, assumes responsibility for providing the means necessary to secure certain defined minimums of well-being, whether in terms of economic security, health, education, or any other standards.

This ensurance of social well-being implies social and economic control by the welfare state, which will regulate consumption as well as production and will further the social interests of groups seen as disadvantaged. Thus a guarantee of collective social care is offered to the members of the society. The state is compelled to move into new social areas and to develop criteria for evaluating its success in achieving these goals. Income is to be redistributed, production assured, an expanding economy planned for; and the "quality of life" is to be defined and monitored, usually in cultural terms.

In all this the social sciences are seen as providing indispensable services for the definition of goals, the design of policy, and the evaluation of programs. The economy is no longer a matter for piecemeal approaches, but must be seen as a whole and planned accordingly. Here, Keynesian economics has played an historic role, both in assuring productivity and buying capacity.

But this does not imply that the welfare state has sailed into ahistorical waters and has reached a state of homeostasis, though some observers might be eager to proclaim this. New problems emerge, and the benefits of the welfare state prove to be rather tentative and unevenly distributed after all. Among other problems, we note that decisions that were once political now seem to be administrative. But the fact that in the welfare state many decisions by-pass the machinery of government does not mean that they are low-level decisions. The expert claims superior scientific knowledge removes these decisions from the political to the administrative spheres.

. . . the democratic system has always rendered homage to the idea that a number of important posts in the government actually called for the possession of more ability than trustworthiness. The waterways, the railways, the post office, the mines, and indeed even the army, are much better served by a sound technical administration than by a management with political responsibilities. As far as the needs of such services were concerned, there was, so to speak, no need for democracy ever to have been invented.

Now in the Welfare State this phenomenon, originally of but marginal importance, suddenly comes right to the centre of the stage. Economic affairs, social affairs, finance, agriculture, social services are all less than before political ministries and more than ever technical ones.[53]

Increasingly, legislative and executive bodies relinquish knowledge and control to the technicians, such that democratic "rituals" acquire less and less meaning, and are even threatened with becoming undignified. For Thoenes, a new form of government has begun to operate as an undercurrent in the welfare state.

What we must particularly bear in mind . . . is the question of whether the Welfare State, as a matter of fact, does not have any real political problems as such, or whether it continues to have them, but in the disguise of technical topics. This is a fundamental question of our time.[54]

Declining and Advancing Elites

The entrepreneur and the engineer are among those whose power and prestige are fading. It is now the technical social scientist, especially the economist, whose star is in the ascendant. It is now the social welfare expert who sees himself as a member of the plan-fulfilling elite, who disdains the distant voice of the politician. His superiority is based on his claim to knowledge of up-to-date data on production, price fluctuations, investment, taxation, employment levels, stock market trends, and the like. "In the democratic system, this knowledge is not supposed to be esoteric," and in fact government agencies publish vast amounts of data at inexpensive prices, normally below cost. But at the same time, few members of the public could even try to master this flood of information. Only the functionary elite possesses the insight for interpreting these facts, for developing ever more abstruse information, obtaining ever more expertly drawn reports, becoming ever more and more esoteric. Thus the function of general edification evaporates, and the "knowledge elite" renders its position ever more unassailable, possessing as it does the appearance of a "scientific cachet." Control over their actions become ever more nominal, by those having less and less knowledge as a basis for forming judgments.

The Apparatus of the Scientific Elite

Having achieved autonomy, the scientific elite will be exhorted to adhere to democratic values, to be impartial in the formulation of programs, to possess objective neutrality; and they will piously claim their faith in these values.

Thus arises the claim—rarely made explicit—that science is an impartial, objective institution, free of the pressures, distortions, and biases that affect all other social institutions, a view that does not hold up under close examination. This is as true for the social sciences as for the hard sciences. They remain dependent on, and are influenced by, specific social structures.

Toward a "Political" Sociology

In his conclusion Thoenes notes that the phenomenon of the functionary elite within the welfare state is "the outcome of an unchaste relationship between politics and science. It is an illegal elite which (no doubt with the best possible intentions) puts a halter on democracy and blindfolds science. The fact that the government calls on a so-called one and indivisible science means that there is brought into politics a monolithic element, with pernicious effects on democracy,"[55] a conclusion in agreement with the observers of American society mentioned above. For Thoenes, the problems with this position are obvious. A science that maintains its pretensions to objectivity will remain unaware of its own foundations, will turn aside from relevant social problems for "pure" scientific research into trivialities, and may lend itself to a blind support of the status quo.

He concludes by calling for an "open sociology."

1. In contrast to that type of sociology which lays stress on the setting up of a closed system of a permanent nature, and which thus might be styled a "closed sociology," the new "political" sociology might be classified as "open," on account of its readiness to undertake fundamental criticism, and of its conviction regarding the incompleteness of history, and its acceptance of the fact that a changing society must lead to a changing sociology, and *vice versa*.

2. Open sociology would have to exist alongside closed sociology. The latter is also necessary, indeed indispensable, as a scientific apparatus. For social work, for town and country planning, for industrial and medical sociology, simply for the purpose of

keeping the machine in working order, a closed sociology is essential. But for such renewal of the machine as from time to time becomes necessary, one would still expect a contribution from the open type of sociology.

3. Open sociology is not a new monolith. Helped along by diverse political outlooks, in the widest sense of the world, it will perpetuate the life of a variety of schools, or give life to them. . . .

4. Open sociology can be kept in being by a relatively small number of sociologists. In contrast to closed sociology, which is concerned with the building up of a machine, it is not dependent on numbers. . . .

5. This does not mean that sociological training can simply jog along the same old road, with the cry: "Open sociology is self-supporting," and get away with it. Training for an open sociology is probably incomparably more difficult than for the closed variety. . . .

6. The fact that the open sociologist does not fit into a convenient set-up, does not simply mean that he has opted for martyrdom. . . .

7. It is essential for an open sociology to have very close contact with non-sociologists . . . particularly in the case of an open political sociology.

8. Contacts with the political world can be obtained through party leaders and scientific bureaus, members of parliament, aldermen and councilors.. . . What is more, it will have to keep an ear open for anything that is whispered in the political demi-monde, a shrinking world certainly . . . but still a very significant one.

9. It will be the special task of an open sociology to put new political country on the map.[56]

This open sociology will be something different from the opposition between the "conflict sociology" of persons like Mills and Dahrendorf on one side, and the structural-functional school of Parsons and Merton. Conflict sociologies can be as rigid and ahistorical as their opponents, the structuralists; conflict sociologies are also limited by the horizons of specific societies and specific historical periods; failing to recognize this, they become simply other variants of "closed" sociology. Thoenes remarks that open sociology may have a great capacity "to administer injections to our present anaemic democracy."

SYSTEMS THINKING AS A HABIT OF MIND

In previous sections we have seen that systems theorists assume the existence of interlocking sets of elements having clearly defined boundaries, purposes, interrelated elements that are generally cooperative. The goals assumed are the goals of the system, not of the actors or of a particular set of actors in a system. Thus the systems goals are given an "objective" character even though it is not possible to identify the goals in a system; they are identified only by the act of assumption or ascription. Yet, given this assumption of goals in a cooperative system, one can raise the question of why and how philosophers of science, and scientists acting as philosophers, posit goals for a system, and if goals are posited, whose goals are they?

A number of possibilities are apparent; though given the level of data we have dealt with, the imputation of goals and motives to people in the analysis above is a speculative phenomenon: The systems theorist may attribute goals to systems, but he does not attribute them to people. In the absence of direct evidence, we are forced to make the transplantation from goals in a system to goals of the persons constructing the system. Among these possibilities we find the following;

Consensus

Systems theorists may assume that the goals of a system, especially here a social system, represent some kind of consensus or social contract. Few of the theorists surveyed above do this, though Talcott Parsons and possibly Easton may do so, at least by implication.

If this is not seen as a consensus, the goals may be a product of the intuitive judgment of the systems theorists who posit goals for systems directly and without inquiry, as a necessity for having a closed system that has purposes and a set of criteria for a series of completed actions. In such instances the goals are those of the systems theorist.

We have seen that much of systems theory derives from the operation of computer systems that have to be logically closed to operate, and that an integrated system of values, goals, cooperation, balance, and equilibrium have to be assumed if there is to be a computer system at all. Thus the personification and anthropomorphization of systems appears to be a by-product of technical necessities. Even if this were true, the systems theorist still makes that imputation, and either his goals or his imagination of what the goals should be become the goals of a system.

However, the basic logic of systems seems to have developed before there were computers; so the development of system goals appears to be part of a "total" or "totalistic" strain of science, and if not of science, of a philosophy of science in which the scientific intellectuals create goals for society that they project into the nature of things. Thus we witness what might be called the objectification and reification of the scientist's attempt to control nature, including human and social nature, by seeing nature as controllable, so that the subjective intentions of the scientific intellectual become reified as a set of "facts," which are treated not as aspirations but as facts.

The Centralization of Society

Since the end of the nineteenth century, the development of centralization of the state, of industry, of the military, of bureaucracy, of all institutions, has led intellectuals in societies undergoing centralization to see the world as a set of responses to the manipulations of central staffs, bureaucrats, and higher officialdom. One could argue, as does Thoenes, that this model of the distributive aspects of society, responding to the central staff of society, has become so natural, so much a part of the perceptive apparatus of scientific intellectuals, that a purified and reified model of an imperfect reality has become the ultimate reality—beyond experience—for scientists who attempt to create, in more pure forms, the systems of thought they unconsciously reflect.

The corresponding model of the nineteenth century, the model of the marketplace created by the Manchester School of liberal economics, had its own teleology, but that teleology was based on the diverse, competitive, conflicting quest for personal pleasure and the avoidance of pain in a widely distributed population of individuals.

If the contemporary systems theorist reifies a model of a centralized and perfected cooperative social system, and if he does this unconsciously, by the very act of so doing, he still adds dimensions. The model of a centralized and bureaucratically operated system as it exists in an empirical world is always imperfect. To the scientist the empirical world is dominated quite frequently by imperfect men, men who do not know enough, who know less than they do, who are not always logical and reasonable, who are often unscientific in their operation, and whose motives may sometimes be corrupt, venal, power-hungry, or "materialistic."

In reifying systems theory and giving it their own (unconsciously) imagined purposes, they change the model, eliminating from it all these contingent imperfections. They give the model a higher unity, a more reasonable set of goals, more altruism and idealism, and they suggest, in their own scientific rationality, a system reflecting their own image of themselves, a technical manipulability that they attribute to science in its ideal sense. Thus the higher motives, the goals of the system, are the higher goals and ideals that they appear to attribute to themselves. (These reflections, to be sure, are all hypothetical.)

SYSTEMS THEORY AND TESTABLE WORK

When systems men do routine technical work for business and industry or for government, they are forced to work within the framework of goals provided by their sponsors; their work under such conditions tends to be more technical, more limited, and more modest in its societal implications. At best, they can do a good job of solving relatively small problems. At worst, if they fail, they fail merely at a technical level. But when they are not subject to the constraints of other people's goals, the claims they make for science can be unlimited. The ideals and purposes they can invest in a system are then grandiose, and since the claims are so far beyond the bounds of testability (no one has as yet turned a whole society over to the systems scientists), they are not subject to the kinds of accountability they must face when asked to undertake relatively simple problems that provide clear-cut results by which their claims could be confirmed or rejected.

At this point, then, science, and the philosophy of science as invested in systems theory, becomes as idealistic, utopian, and impractical as is any non-scientific teleological system, including magic, and is perhaps even less subject to testing and accountability.

THE SCIENTIST-KING

If this is the impulse of systems theory when it goes beyond specific, limited, testable work, the roots of this impulse are easy to see. Since the very beginning of intellectual self-awareness, the proponents of reason, of philosophy, and of science, have asked the powers of the world to surrender their control of the world to them. With their tools of reason, conceptualization, and science, they have assumed they could solve all problems. Plato was perhaps the first of these, yet he can stand as a paradigmatic figure for hundreds of intellectuals who felt that the discovery of a new intellectual tool was the basis for a claim to power, among them men such as Macchiavelli, Vico, Comte, St. Simon, Kluckhohn, and countless others.

The intellectual—scientific, sociological, philosophical, political—automatically assumes that knowledge, because he values it, is pure, and that the program and goals of science are so evidently valuable in themselves that once they demonstrate the superiority of their knowledge, all others will surrender their power to them on the basis of their superiority of knowledge.

Systems theorists fail to recognize that knowledge is not power and that others will still cling to economic or political or social motives as much as some cling to scientific motives. Moreover, they fail to realize that science, for all these others, can be instrumental—a means, not an end—and therefore they fail to realize that science does not serve science, but serves other ends in the external world, even while scientists want to believe that science serves science or scientists.

Finally, in asserting that science serves science, they attempt to convert the total world collectively into the realm of science. In so doing, by giving to the system their purposes, ideals, motives, and intentions, they are making a claim to power. The world as a gigantic laboratory is their world, and their conception of systems is a world they can manipulate in order to make science. In manipulating that world they are not simply making science, they are ruling the world.

They are not monsters out of science fiction or an immediate threat. So far as we know, they suffer the relatively innocent delusions of men who do not know their limits, but who have as yet been unable to escape the limits that are part of their historical and empirical situation. That is, nobody grants them the control that is implicit in the systems they set up for themselves. Those in control use science but have so far not abandoned their own perspectives or their own control. The systems constructed by the systems theorists are at best playthings of scientists, or perhaps sales presentations—very dignified ones, to be sure—to scientifically less-sophisticated politicians, administrators, and businessmen, who for short periods of time may adopt the systems of science, as long as they produce immediately practical results, judged from the standpoint of criteria they determine.

When the scientists, perhaps because of their excessive claims—excessive because they are based on their idealization of their functions—mess things up, they are replaced, sometimes, by other scientists making other sets of claims based on different techniques, methods, and perhaps even different "systems."

To conclude this discussion of modern tyranny and the part intellectuals play in it, one might ask what in the end is the principle of authoritarianism, and in what specific form it appears nowadays.

The simplest reply seems to be that the authoritarian principle is inherent in the very fact of placing oneself, as regards social and political problems, at the point of view of the ensemble, the totality, of the necessary . . . congruence of the parts and the efficient functioning of the whole. In fact, the preoccupation with the totality implies the idea that human society is an organism whose laws are essentially known and, by implying that, it also implies the idea that one can, indeed one must, modify it from on high by means of more or less violent external interventions.[57]

NOTES

1. Harold D. Lasswell, *World Politics and Personal Insecurity* (New York: Free Press, 1965; first published, 1935), p. 181.
2. Ervin Laszlo's claim. See Ch. 6 above.
3. H. A. Hodges, *Wilhelm Dilthey—An Introduction* (London: Routledge and Kegan Paul; New York: Howard Fertig, 1969; first published 1944), p. viii. This discussion is much indebted to John Lukacs, *Historical Consciousness, or the Remembered Past* (New York: Harper & Row, 1968), Ch. 7, "History and Physics, or the end of the Modern Age."
4. Jose Ortega y Gasset, "A Chapter from the History of Ideas—Wilhelm Dilthey and the Idea of Life," in *Concord and Liberty*, trans. by Helene Weyl (New York: Norton, 1946; reprinted 1963).
5. Werner Heisenberg, "The Role of Modern Physics in the Present Development of Human Thinking," Ch. 11 of *Physics and Philosophy—The Revolution in Modern Science* (New York: Harper & Row, 1958; reprinted 1962), p. 200. Emphasis added. These essays were originally delivered as lectures in 1955.
6. John Lukacs, *op. cit.*, pp. 278ff.
7. Heisenberg, *op. cit.*, pp. 102–105. Emphasis added.
8. *Ibid.*, pp. 106–107. Emphasis added.
9. *Ibid.*, pp. 40–41. See also Lukacs, *op. cit.*, pp. 283–285.
10. *Ibid.*, pp. 182–185.
11. Ashby touches upon this problem, in order to dismiss it:

> It will be further assumed (except where the contrary is stated explicitly) that the functioning units of the nervous system, and of the environment, behave in a determinate way. By this I mean that each part, if in a particular state internally and affected by particular conditions externally, will behave in one way only. (This is the determinacy shown, for instance, by the relays and other parts of a telephone exchange.) It should be noticed that we are not assuming that the *ultimate* units are determinate, for these are atoms, which are known to behave in an essentially indeterminate way; what we shall assume is that the *significant* unit is determinate. The significant unit (e.g., the relay, the current of several milliamperes, the neuron) is usually of a size much larger than the atomic so that only the average property of many atoms is significant. These averages are often determinate in their behaviour, and it is to these averages that our assumption applies.
>
> The question whether the nervous system is composed of parts that are determinate or stochastic has not yet been answered . . . we shall suppose that they are determinate. . . . But we need not pre-judge the issue; [this] book is an attempt to follow the assumption of determinacy wherever it leads. When it leads to obvious error will be time to question its validity.
>
> W. Ross Ashby, *Design for a Brain*, pp. 9–10. Thus Ashby fails to draw the conclusions suggested by Heisenberg.

12. Owen Barfield, *What Coleridge Thought* (Middletown, Conn.: Wesleyan University Press, 1971), p. 205.
13. See Ervin Laszlo, *Introduction to Systems Philosophy* pp. 242–249.
14. George Lichtheim, *The Concept of Ideology and Other Essays* (New York: Vintage, 1967), pp. 3–46. This discussion follows Lichtheim closely on some but not all points.
15. *Ibid.*, p. 3.

16. *Ibid.,* p. 8.
17. Lichtheim, *op. cit.,* p. 9, citing Helvetius, *De l'Esprit,* (1758).
18. *Ibid.,* p. 10.
19. *Ibid.,* p. 19 and pp. 21–22.
20. *Ibid.,* p. 32.
21. Karl Mannheim, *Ideology and Utopia—An Introduction to the Sociology of Knowledge,* trans. by Louis Wirth and Edward Shils (New York: Harcourt Brace, 1965). The American version contains material not included in the original version.
22. *Op. cit.,* pp. 39–40.
23. *Ibid.,* p. 40.
24. *Ibid.,* p. 154–155.
25. *Ibid.,* pp. 160–161.
26. It might also be added that Mannheim himself showed enough positivist elements to render him susceptible to such claims for the intellectuals; his later work, on planning in a democratic society, showed an affinity for the collectivist potentials within systems theory; like the systems theorists, he hoped that a democratic "spirit" would animate the bureaucratic body. Somehow, he seemed to hope, the planners would be animated enough by a "democratic" spirit not to seize and use the power positions they would hold. Cf., e.g., Mannheim's *Man and Society in an Age of Reconstruction,* Part IV, Section IX: "The problem of Transforming Man;" also, his entire *Freedom, Power and Democratic Planning,* and, in his *Essays on Sociology and Social Psychology,* Part IV, consisting of four lectures under the title, "Planned Society and the Problem of Human Personality: A Sociological Analysis." For Mannheim, the alternative to liberal (i.e., laissez-faire) society was to be planned society, not totalitarian society. The problem of a possible totalitarianism of the planners was left unsolved.
27. Howard Becker and Harry Elmer Barnes, *Social Thought from Lore to Science* (New York: Dover, 1961; originally published 1938), Vol. I, pp. 80ff., *et passim,* Vol. II, pp. 679ff., on "The Biologizing of Social Theory" in Comte, Spencer, Paul von Lilienfeld, and others. See also Ralph Henry Bowen, *German Theories of the Corporative State, with Special Reference to the Period 1870–1918* (New York: Whittlesey House, 1947), Frank E. Manuel, *The Prophets of Paris* (New York: Harper & Row, 1962), George Iggers, trans., *The Doctrine of Saint-Simon—An Exposition* (New York: Schocken, 1962, first published 1958).
28. John Lukacs, *Historical Consciousness,* pp. 39ff: "Our Interregnal Condition:" "What is unique in the twentieth century is that we *think* that we live in an intellectually revolutionary age— when, so far as ideas go, this is not really so: during our democratic interregnum, all superficial impressions and the extraordinary rapidity of external communications notwithstanding, the movement of ideas has been remarkably slow. . . . For, with few exceptions we have been moving on by the momentum of nineteenth-century ideas of 'progress'—which is why the movement of ideas has been so slow" (p. 41).
29. Alexis de Tocqueville, *Democracy in America,* trans. by George Lawrence, ed. by J. P. Mayer (Garden City, N.Y.: Doubleday, 1969); especially Vol. 2, *passim.* William Kuhns, *The Post-Industrial Prophets* (New York: McKay, 1971; reprint, New York: Harper & Row, 1973); James Martin and Adrian R. D. Normal, *The Computerized Society* (Englewood Cliffs, N.J.: Prentice-Hall, 1970); Brian Rothery, *The Myth of the Computer* (London: Business Books, 1971); Robert Boguslaw, *The New Utopians—a Study of System Design and Social Change* (Englewood Cliffs, N.J.: Prentice-Hall, 1965).
30. H. L. Nieburg, *In the Name of Science,* rev. ed. (Chicago: Quadrangle, 1970); for a summary of this work, see Kleinberg, *American Society in the Postindustrial Age,* Ch. 9 (Columbus, Ohio: Charles E. Merrill, 1973).
31. Theodore J. Lowi, *The End of Liberalism—Ideology, Policy, and the Crisis of Public Authority* (New York: Norton, 1969).

32. Kleinberg, *op. cit.*, p. 162.

33. Joseph Bensman and Arthur Vidich, *The New American Society* (Chicago: Quadrangle, 1971), Chs. 5 and 6.

34. Kleinberg, *op. cit.*, p. 162.

35. Robert A. Dahl and Charles E. Lindblom, *Politics, Economics, and Welfare* (New York: Harper & Row, 1953), p. 256; cited in Kleinberg, *op. cit.*

36. Fritz Morstein Marx, *The Administrative State—An Introduction to Bureaucracy* (Chicago: The University of Chicago Press, 1957), pp. 20–21.

37. *Ibid.*, p. 32.

38. *Ibid.*, Ch. 4. This summary omits much concrete historical detail.

39. An awareness of bureaucratic authoritariansim is of course as old as Weber; further, Friedrich Hayek's *The Road to Serfdom* (Chicago: The University of Chicago Press, 1944) expressed the same concerns more systematically and programmatically some time ago. What we are concerned with here is the empirical analysis of these traits emerging in the new administrative state, which is part of a "democratic" polity.

40. Ralf Dahrendorf, *Class and Class Conflict in Industrial Society* (Stanford, Calif.: Stanford University Press, 1959), p. 303.

41. Ralf Dahrendorf, "Market and Plan—Two Types of Rationality," in *Essays in the Theory of Society* (Stanford, Calif.: University of Stanford Press, 1968), p. 228. Dahrendorf does not consider "plan rationality" to be compatible with human freedom.

42. Lee Rainwater and William L. Yancey, *The Moynihan Report and the Politics of Controversy* (Cambridge: M.I.T. Press, 1967), Chs. 12 and 13, and *passim.*

43. Daniel P. Moynihan, *Maximum Feasible Misunderstanding—Community Action in the War on Poverty* (New York: Free Press, 1970), "Introduction to the Paperback Edition," pp. xvii–xix. The views of Krause are quoted from Elliott A. Krause, "Functions of a Bureaucratic Ideology: 'Citizen Participation,'" in *Social Problems*, Vol. 16, No. 2 (Fall 1968), pp. 129–143.

44. Moynihan, *Maximum Feasible Misunderstanding*, pp. xxii–xxiii.

45. Daniel P. Moynihan, *The Politics of a Guaranteed Income—The Nixon Administration and the Family Assistance Plan* (New York: Vintage, 1973), pp. 304ff., *et passim.*

46. Piet Thoenes, *The Elite in the Welfare State*, trans. by J. E. Brigham, ed. by J. A. Banks (London: Faber & Faber, 1966; originally published as *De Elite in de Verzorgingstaat* (Leiden, Holland; H. E. Stenfert Kroese, 1962).

47. *Ibid.*, p. 25.

48. *Ibid.*, p. 45.

49. *Ibid.*, p. 64.

50. *Ibid.*, p. 90.

51. *Ibid.*, p. 115. Compare the discussion in Bensman and Vidich's *The New American Society*, Chs. 5, "The Coordination of Organizations" and 6 "Coordination and Competition among Elites.")

52. *Ibid.*, p. 127.

53. *Ibid.*, p. 166.

54. *Ibid.*, p. 168.

55. *Ibid.*, p. 217.

56. *Ibid.*, pp. 221ff.

57. Nicola Chiaromonte, "On Modern Tyranny—A Critique of Western Intellectuals," *Dissent*, March–April 1969, pp. 137–150; reprinted in *The Worm of Consciousness and Other Essays*, (New York, Harvest Books, 1976).

Index of Names

Index of Topics

Index of Books, Articles, and Publishings